B 1 499473 9

Education and Neuroscience

This book brings together contributions from experts in education and the neurosciences at the forefront of interdisciplinary research efforts to build bridges between these two areas. It includes chapters that critically consider the relevance of the sciences of mind and brain to educational areas such as reading, mathematics, music and creativity. These chapters support the notion that there is, indeed, much to justify the growing interest of educators in neuroscience. However, this enthusiasm also brings dangers with it, as evidenced by the proliferation of neuromyths within schools and colleges. For this reason, it also reviews some of the more prominent misconceptions, as well as exploring how educational understanding can be constructed in the future to include concepts from neuroscience more judiciously.

This book will be of interest to educators, policymakers and scientists seeking fresh perspectives on how we learn.

This book was previously published as a special issue of *Educational Research*, a journal of the National Foundation for Educational Research (NFER).

Paul Howard-Jones specialises in interdisciplinary research involving neuroscience and education, publishing in education, philosophy, psychology and neuroscience. His research, whether using educational, psychological or neuroscientific methods, is grounded by considerable past experience in the training and professional development of teachers. In 2005-2006, he coordinated the UK's ESRC seminar series on Neuroscience and Education, authoring the popular commentary that arose from it. He is also the author of the recently published *Introducing Neuroeducational Research: Neuroscience, education and the brain from contexts to practice* (2009). He is a passionate contributor to the general debate around neuroscience and education in educational, scientific and public arenas, but his more specific research interests include creativity, educational technology and learning games. He co-ordinates the NeuroEducational Research Network (NEnet) at the Graduate School of Education, University of Bristol, UK.

Education and Neuroscience

Evidence, Theory and Practical Application

Edited by Paul Howard-Jones

Routledge
Taylor & Francis Group

LONDON AND NEW YORK

First published 2010
by Routledge
2 Park Square, Milton Park, Abingdon, Oxon, OX14 4RN

Simultaneously published in the USA and Canada
by Routledge
270 Madison Avenue, New York, NY 10016

Routledge is an imprint of the Taylor & Francis Group, an informa business

© 2010 NFER

Typeset in Times by Value Chain, India
Printed and bound in Great Britain by TJI Digital, Padstow, Cornwall

British Library Cataloguing in Publication Data
A catalogue record for this book is available from the British Library

ISBN10: 0-415-56496-4
ISBN13: 978-0-415-56496-0

CONTENTS

INTRODUCTION

This book focuses on a new and exciting area of interdisciplinary research emerging at the interface between neuroscience and education. On the face of it, the need for such research appears obvious. If we are learning so much about the brain, surely this can help us improve education? Arguably, you might say teachers are the only professionals charged with the daily development of brain function, and one scientist (Koizumi 2004) has even suggested that education might be defined as a 'nurturing of the brain'.

Between 2005 and 2006, the ESRC-TLRP seminar series 'Collaborative Frameworks in Neuroscience and Education' brought together over 400 teachers, neuroscientists, psychologists and policy-makers to discuss the potential for collaborative work that might lead to improved educational and neuroscientific understanding. This book brings together and examines many of the issues and opportunities highlighted by the seminar series, by drilling down into just a few of many topics touched upon in the associated commentary (Howard-Jones 2007). Public interest in neuroscience and education has been considerable, as reflected by media coverage and the unprecedented accessing of the seminar series commentary (more than 110,000 downloads in the first 6 months). But this enthusiasm also brings with it dangers, as evidenced by the long-running success of entrepreneurs in constructing and promoting unscientific and unevaluated 'brain-based' pedagogy. It is appropriate, therefore, that the book should begin with a provocative article by John Geake scrutinising some of the most popular ideas about the brain to be found in today's classroom. Geake examines ideas about using 10% of our brain, left- and right-brain thinking, Visual Auditory Kinaesthetic (VAK) learning styles and multiple intelligences (as they are often interpreted in education). All these concepts are appealing in their simplicity and may resonate with some educational viewpoints. However, while they have usually been inspired by something related to neuroscience, any scientific basis has been so seriously misinterpreted, over-interpreted and/or misapplied that they are classified here as 'neuromyths'. By pulling the fallacies apart, Geake provides a convincing argument for developing a mutually comprehensible language and genuine interdisciplinary dialogue, in order to avoid these and future pitfalls.

By the end of Geake's article, the reader may be left asking: 'if these were just neuromyths, then what are we learning about education and the brain that is genuine and important?' In considerable contrast to the concepts critiqued by Geake, the next four chapters in this book provide fascinating examples of how brain research is revealing insights about learning that have genuine implications for educational practice. Usha Goswami reviews what we know about the core neural systems involved with learning to read and the biological basis of developmental dyslexia. In so doing, Goswami highlights the complementary role of different brain imaging techniques in constructing this understanding, suggesting that future studies should draw selectively upon these techniques and approach difficult issues of causation through longitudinal design. As in other areas of educational interest, there are relatively few developmental studies of the brain and reading at present, with most neuroimaging studies involving only adult participants. This is partly due to the inappropriateness and difficulty of using some imaging methods with children, such that child-friendly techniques like EEG appear set to develop a special significance within neuroscience and education. Goswami also draws attention to the exciting possibility that EEG may be helpful in identifying early neural markers of risk for developmental dyslexia in infants.

Goswami demonstrates convincingly that existing studies suggest dyslexia is associated with an under-activation of key networks involved with reading, but also notes the shift during normal development of the brain areas involved with language as they become increasingly left-lateralised in most readers. This theme of shifting activity during development is picked up in the third paper by Sashank Varma and Daniel Schwartz. The essential role of cognition as a bridge between neuroscience and education is well known, but Varma and Schwartz ask the important question of how this bridge should be best conceptualised. Theoretical approaches are divided into an 'area focus' that understands a cognitive competency in terms of one brain area, and a network focus that explains it as a product of collaborative processing among many brain areas. It is clear from Varma and Schwartz's analysis that this difference in scientific theorising can have direct implications for the interpretation of results in terms of educational practice. An approach that highlights qualitative shifts in underlying networks also prompts educational ideas about how best to design and schedule support for such shifts, with potential implications for curriculum and teaching strategy.

Liane Kaufmann, in her paper, provides a further example that illustrates the importance of a network focus approach, and draws attention to developmental differences in the neural mechanisms linking numerical processing and the use of fingers. Before focusing on this particular aspect of early numeracy, she makes the point that single-deficit models, corresponding to Varma and Schwartz's 'area focus' approach, have probably been more popular in the literature because they are simpler and more testable than dynamic, process-oriented and multiple-deficit approaches. This is one example of why caution must be applied when considering how ideas from education, which often embrace the situated and complex nature of learning, can be combined with concepts arising from natural science perspectives that embrace parsimony. Kaufmann, as Varma and Schwartz, provides a coherent argument why such a more complex approach is required and emphases that, despite the additional difficulties this provides for scientific investigators, such complex models are more likely to lead to meaningful conceptual links between mind, brain and education. Kaufmann's own research in studying numerical development emphasises the value of neuroimaging techniques, especially when they reveal differences that are not reflected in performance. Once again, a shifting in underlying neural networks during development is revealed, and this leads to the suggestion that fingers may have a special role as concrete embodied tokens for representing number magnitude. Finally, Kaufman tentatively proposes the explicit incorporation of finger-use in intervention programs for dyscalculia as a potentially interesting area for future research.

While a body of research now exists linking numeracy and literacy with brain function, the area of music and the brain is only just starting out. For that reason, the contribution by Lauren Stewart and Aaron Williamon is particularly welcome, and provides a pioneering review of literature identifying a new and potentially very rewarding area for research. Consideration of the neural basis of music also focuses the reader on issues of cultural context and individual differences since, as Stewart and Williamon point out, music does not exist in the outside world, but is made sense of by multiple brain areas across both hemispheres. While all brains are supremely adapted for making links and seeking patterns and meaning, the way this is achieved with music varies widely between cultures and also between individuals within the same culture. Stewart and Williamon's paper also dissuades us from any notion that biology is destiny. They review evidence for natural abilities with a genetic origin, but make the point that these do not develop by biological maturation alone, but require stimulation through practice and learning. Deliberate practice is the prime predictor of changes in performance and the neural processes associated with them. Even after a short period of training on musical notation, learners show behavioural and biological evidence of processes becoming automatic, with evidence from other (non-musical) studies showing how changes in function can, in the longer term and across the life-span, become associated with structural changes. This undermines the popular notion that differences at a brain level must indicate some type of biologically determined issue. Stewart and Williamon review many more insights, including issues of memorising music, and the apparently valuable role of neurofeedback in enhancing performance. In furthering research that combines neuroscience with music, or any other educational area, I would strongly support Stewart and Williamon's proposal that integrated neuroscientific, behavioural and observational studies are needed.

However, mindful of our first paper by Geake, what is to stop the many insights and ideas emerging from these areas of research in neuroscience and education becoming just another set of 'neuromyths'? Even VAK probably began with a scientifically observable piece of evidence – e.g., that we exhibit individual preferences in how we learn. Somewhere along the line(s) of communication, this commonsense notion mutated into the need for children to be labelled V, A or K and for teaching styles to be differentiated accordingly. Scientific findings are never likely to directly translate into lesson plans, so what sort of communication processes would help produce pedagogical concepts that are more educationally useful and scientifically credible than the present neuromythology? In the final contribution to this book Howard-Jones, Winfield and Crimmins explore this important question, by reporting on an interdisciplinary attempt to co-construct pedagogical ideas spanning neuroscience and education. The context was drama education, but the findings echo some of the general issues highlighted by other authors in this book. In particular, it highlights the ease with which neuromyths can be generated but also the immense, and mostly untapped, relevance of our understanding of mind and brain to education. Trainee drama teachers co-constructed their ideas with a research team possessing educational and appropriate scientific expertise. Interestingly, trainees passed through a series of observable stages in their approach. After initial enthusiasm and the generation of more neuromyths, a daunting sense of complexity arose before participants started focusing on models of cognition mutually informed by neuroimaging studies, and then began reflecting upon their practice with a new sense of depth and insight. Was the process we observed here in any sense reflective of the broader processes whereby neuroscience is beginning to influence educational thought? If so, I would speculate that we may now be at that daunting stage of realising that neuroscience has no ready-made prescriptive answers for education. Instead, through careful integration of scientific insights with educational expertise and understanding, what we are learning about the brain and the mind is promising to enrich educational perspectives in more subtle, meaningful and valuable ways.

Paul A. Howard-Jones
Graduate School of Education, University of Bristol, UK

References

Koizumi, H. 2004. The concept of 'developing the brain': A new natural science for learning and education. *Brain and Development* 26: 434–41.

Howard-Jones, P.A. 2007. Neuroscience and education: Issues and opportunities. In *TLRP commentary*. London: TLRP. http://www.tlrp.org/pub/commentaries.html.

Neuromythologies in education

John Geake

Oxford Brookes University, Oxford, UK

Background: Many popular educational programmes claim to be 'brain-based', despite pleas from the neuroscience community that these neuromyths do not have a basis in scientific evidence about the brain.

Purpose: The main aim of this paper is to examine several of the most popular neuromyths in the light of the relevant neuroscientific and educational evidence. Examples of neuromyths include: 10% brain usage, left- and right-brained thinking, VAK learning styles and multiple intelligences

Sources of evidence: The basis for the argument put forward includes a literature review of relevant cognitive neuroscientific studies, often involving neuroimaging, together with several comprehensive education reviews of the brain-based approaches under scrutiny.

Main argument: The main elements of the argument are as follows. We use most of our brains most of the time, not some restricted 10% brain usage. This is because our brains are densely interconnected, and we exploit this interconnectivity to enable our primitively evolved primate brains to live in our complex modern human world. Although brain imaging delineates areas of higher (and lower) activation in response to particular tasks, thinking involves coordinated interconnectivity from both sides of the brain, not separate left- and right-brained thinking. High intelligence requires higher levels of inter-hemispheric and other connected activity. The brain's interconnectivity includes the senses, especially vision and hearing. We do not learn by one sense alone, hence VAK learning styles do not reflect how our brains actually learn, nor the individual differences we observe in classrooms. Neuroimaging studies do not support multiple intelligences; in fact, the opposite is true. Through the activity of its frontal cortices, among other areas, the human brain seems to operate with general intelligence, applied to multiple areas of endeavour. Studies of educational effectiveness of applying any of these ideas in the classroom have failed to find any educational benefits.

Conclusions: The main conclusions arising from the argument are that teachers should seek independent scientific validation before adopting brain-based products in their classrooms. A more sceptical approach to educational panaceas could contribute to an enhanced professionalism of the field.

Introduction

Neuromythologies are those popular accounts of brain functioning, which often appear within so-called 'brain-based' educational applications. They could be categorised into neuromyths where more is better: 'If we can get more of the brain to "light up", then learning will improve...', and neuromyths where specificity is better: 'If we concentrate teaching on the "lit-up" brain areas then learning will improve . . .'. Prominent examples of neuromythologies of the former include: the 10% myth, that we only use 10% of our brain; multiple intelligences; and Brain Gym. Prominent examples of neuromytholgies of the latter include: left- and right-brained thinking; VAK (visual, auditory and kinaesthetic) learning styles; and water as brain food. Characteristically, the evidential basis of these schemes does not lie in cognitive neuroscience, but rather with the various enthusiastic promoters; in fact, sometimes the scientific evidence flatly contradicts the brain-based claims.

The assumption here is that educational practices which claim to be concomitant with the workings of the brain should, in fact, be so, at least to the extent that the scientific jury can ever be conclusive (Blakemore and Frith 2005). A counter-argument might be posed that the ultimate criterion is pragmatic, not evidential, and if it works in the classroom who cares if it seems scientifically untenable. For this author, basing education on scientific evidence is the hallmark of sound professional practice, and should be encouraged within the educational profession wherever possible. The counter-argument only serves to undermine the professionalism of teachers, and so should be resisted.

This is not to say that there is not a glimmer of truth embedded within various neuromyths. Usually their origins do lie in valid scientific research; it is just that the extrapolations go well beyond the data, especially in transfer out of the laboratory and into the classroom (Howard-Jones 2007). For example, there is plenty of evidence that cognitive function benefits from cardiovascular fitness; hence, general exercise is good for the brain in general (Blakemore and Frith 2005). But this does not mean that pressing particular spots on one's body, as per Brain Gym, will enhance the activation of particular areas in the brain. As another example, there are undoubtedly individual differences in perceptual acuities which are modality based, and include visual, auditory and kinaesthetic sensations (although smell and taste are more notable), but this does not mean that learning is restricted to, or even necessarily associated with, one's superior sense. All of us have areas of ability in which we perform better than others, especially as we grow older and spend more time on one rather than another. Consequently, a school curriculum which offers multiple opportunities is commendable, but this does not necessarily depend on there being multiple intelligences within each child which fortuitously map on to the various areas of curriculum. General cognitive ability could just as well play an important role in learning outcomes across the board.

The generation of such neuromythologies and possible reasons for their widespread acceptance has become a matter for investigation itself. In particular, the phenomenon of their widespread and largely uncritical acceptance in education raises several questions: why has this happened?; what might this suggest about the capacity for the education profession to engage in professional reflection on complex scientific evidence? And one cannot help but wonder about the extent to which political pressure for endless improvement in standardised test scores, publicised via school league tables, drives teachers to adopt a one-size-fits-all, brain-based life-raft when their daily classroom experience is replete with children's individual differences.

To gather some data about these issues, Pickering and Howard-Jones (2007) surveyed nearly 200 teachers either attending an education and brain conference in the UK (one brain based, the other academic) or contributing to an OECD website internationally. All respondents were enthusiastic about the prospects of neuroscience informing teaching practice, particularly for pedagogy, but less so for curriculum design. Moreover, despite a prevailing ethos of pragmatism (notably with the brain-based conference attendees), it was generally conceded that the role of neuroscientists was to be professionally informative rather than prescriptive. This, in turn, points to the critical necessity for a mutually comprehensible language with which neuroscientists and educators can engage in a genuine interdisciplinary dialogue.

The American Nobel Laureate physicist Richard Feynman, in one of his more famous graduation addresses at Caltech, warned his audience of young science graduates about 'cargo cult science' (Feynman 1974). His point was that, while it might accord with 'human nature' to engage in wishful thinking, good scientists have to learn not to fool themselves. Feynman's warning could well be applied to the myriad 'brain-based' strategies that pervade current educational thinking. Whereas it is commonly stated in such schemes that the brain is the most complex object in the universe (although how this could possibly be verified remains unexplained), this assumption is then completely ignored in proposing a pedagogy based on the simplest of analyses – e.g., in the brain there are two hemispheres, left and right, therefore there are two kinds of thinking: of-the-left-brain and of-the-right-brain, and therefore there are only two kinds of teaching necessary: for-the-left-brain and for-the-right-brain. Not a very exciting universe where the most complex object has only two states! And not, fortunately, the universe in which we exist, where the complexity of the human brain has been the focus of intense investigation for over a century, but particularly over the past two decades, thanks to the invention of neuroimaging technologies.

The resulting neuroimages – brains with brightly coloured areas – are disarmingly simple, and seem to fit with a commonsense view of the brain as having localised specialist functions which enable us to do the various things we do. But such apparent simplicity is generated out of considerable complexity. In functional magnetic resonance imaging (fMRI), for example, the images are the end-result of many years' work on understanding the quantum mechanics of nuclear magnetic resonance phenomena, the development of the engineering of superconducting magnets, the application of inverse fast Fourier transforms to large data sets and the refinement of high-speed computing hardware and software to analyse large data sets across multiple parameters. The neuroimaging picture is undoubtedly worth the proverbial thousand words, but the scientist's words can be quite different from those of the layperson.

A crucial point that most of the media overlook, or ignore, is that neuroimaging data are statistical. The coloured blobs on brain maps representing areas of significant activation (so-called 'lighting up') are like the peaks of sub-oceanic mountains which rise above sea level, in neuroimaging, how much or how little activation (sea level) to reveal being determined by the researcher in setting a suitable level of statistical threshold. In fact, the most challenging aspect of most neuroimaging experimental design is to determine suitable control conditions to highlight a particular area of experimental interest and thus avoid showing how most of the brain is involved in most cognitive tasks. So, in a classroom it would be quite silly to think that only a small portion of pupils' brains are involved in a task, just because a small area of brain activity was reported in a neuroimaging study of a similar task (Geake 2006). Neuroscience is a laboratory-based endeavour. Even with the best of intentions, extrapolations from the lab to the classroom need to be made with considerable caution (Howard-Jones 2007). As Nobel Laureate Charles Sherrington (1938, 181) warned in Oxford some 70 years ago: 'To suppose the roof-brain consists of point to point centres identified each with a particular item of intelligent concrete behaviour is a scheme over simplified and to be abandoned.' In other words, we have to be very wary of oversimplifications of the neuro-level of description in seeking applications at the cognitive or behavioural levels.

The central characteristic of brain function which generates its complexity is neural functional interconnectivity. There are common brain functions for all acts of intelligence, especially those involved in school learning (Geake in press). These interconnected brain functions (and implicated brain areas) include:

- Working memory (lateral frontal cortex);
- Long-term memory (hippocampus and other cortical areas);
- Decision-making (orbitofrontal cortex);
- Emotional mediation (limbic subcortex and associated frontal areas);
- Sequencing of symbolic representation (fusiform gyrus and temporal lobes);
- Conceptual interrelationships (parietal lobe);
- Conceptual and motor rehearsal (cerebellum).

This parallel interconnected functioning is occurring all the time our brains are alive. Importantly, these neural contributions to intelligence are necessary for all school subjects, and all other aspects of cognition. Creative thinking would not be possible without our extensive neural interconnectivity (Geake and Dobson 2005). Moreover, there are no individual modules in the brain which correspond directly to the school curriculum (Geake 2006). Cerebral interconnectivity is necessary for all domain-specific learning, from music to maths to history to French as a second language. Neuromyths typically ignore such interconnectivity in their pursuit of simplicity. Steve Mithen (2005) argues that it was a characteristic of the Neanderthal brain that it was not well interconnected. This could explain the curious stasis of Neanderthal culture over several hundred thousand years, and the even more curious fact that Neanderthal culture was rapidly out-competed by our physically less robust Cro-Magnon forebears, whose brains, Mithen argues, had evolved to become well interconnected.

Multiple intelligences

Highly evolved cerebral interconnectedness has implications for any brain-based justification of the widely promoted model of multiple intelligences (MI). Gardner (1993) divided human cognitive abilities into seven intelligences: logic-mathematics, verbal, interpersonal, spatial, music, movement and intrapersonal. Some 2500 years earlier, Plato recommended that a balanced curriculum have the following six subjects: logic, rhetoric, arithmetic, geometry-astronomy, music and dance-physical. For philosopher-kings, additionally, *meditation* was recommended. Clearly MI is nothing new: Gardner has just recycled Plato. But although such a curriculum scheme is long-standing, it doesn't mean that our brains think about these areas completely independently from one another. Each MI requires sensory information processing, memory, language, and so on. Rather, this just demonstrates Sherrington's point that the way the brain goes about dividing its labours is quite separate from how we see such divisions on the outside, so to speak. In other words, there are no multiple intelligences, but rather, it is argued, multiple applications of the same multifaceted intelligence.

Whereas undoubtedly there are large individual differences in subject-specific abilities, the evidence which conflicts with a multiple intelligences interpretation of brain function is that these subject-specific abilities are positively correlated, as shown by Carroll (1993) in his large meta-analysis. Such a pervasive correlation between different abilities is conceptualised as general intelligence, g. The existence of g not only suggests that the same brain modules are likely to be involved in many different abilities, but that their functional connectivity is of paramount importance. In fact, the main thrust of research in cognitive neuroscience in the next decade will be the mapping of functional connectivity, that is how functional modules transfer information, anatomically, bio-chemically, bio-electrically, rhythmically, synchronistically, and so on. A recent study along these lines sought evidence for neural correlates of general intelligence – i.e., where and how does the brain generate measures of general intelligence? Duncan et al. (2000) found a common brain involvement, in the frontal cortex of adult subjects, on both spatial and verbal IQ tests. A further meta-analysis of 20 neuroimaging studies involving language, logic, mathematics and memory showed that the same frontal cortical areas were involved (Duncan 2001). It seems unlikely that these intelligences are independent if the same part of the brain is common to all. This point is elaborated in a recent critique of MI (Waterhouse 2006, 213).

> The human brain is unlikely to function via Gardner's multiple intelligences. Taken together the evidence for the intercorrelations of subskills of IQ measures, the evidence for a shared set of genes associated with mathematics, reading, and *g*, and the evidence for shared and overlapping 'what is it?' and 'where is it?' neural processing pathways, and shared neural pathways for language, music, motor skills, and emotions suggest that it is unlikely that each of Gardner's intelligences could operate 'via a different set of neural mechanisms' [as Gardner claims].

To explain how those same pathways support high-level general intelligence across so many different cognitive areas, Duncan (2001, 824) suggested that: 'neurons in selected frontal regions adapt their properties to code information of relevance to current behaviour, pruning away ... all that is currently task-irrelevant.' So, underlying our specific abilities is adaptive brain functioning. In support of this idea of an adapting brain, Dehaene and his colleagues have proposed a dynamic model of brain functioning in which these frontal adaptive neurons coordinate the myriad inputs from our perceptual modules from all over the brain, and continually assess the relative importance of these inputs such that from time to time, a thought becomes conscious; it literally 'comes to mind' (Dehaene, Kerszberg, and Changeux 1998). It could be predicted, then, that deliberate attempts to restrict intelligence within classrooms according to MI theory would not promote children's learning, and it could be noted in passing that one of the 'independent consultants' who advocates brain-based learning strategies acknowledges teachers' frustration with the lack of long-term impact of applying MI theory (Beere 2006).

10% usage

None of the above implies that *g* is all that there is to intelligence – quite the opposite. With its population age-norming, IQ might be a convenient surrogate for intelligence in the laboratory, but not even the most resolute empiricist would claim that IQ captures all of the variance in cognitive abilities. Rather, intelligence in all its manifestations illustrates the underlying dynamic complexity of its generative neural processes, with emphasis on 'dynamic'. There is overwhelming evidence that the brain is perpetually busy, and that even when any of our brain cells are not involved in processing some information, they still fire randomly. As an organ which has evolved not to know what is going to happen next, such constant activity keeps our brain in a state of readiness. Consequently, the neuromyth that 'We only use 10% of our brains' could not be more in error. The absurdity has been pointed out by Beyerstein (2004): evolution does not produce excess, much less 90% excess. In the millions of studies of the brain, no one has ever found an unused portion of the brain!

It is unfortunate that teachers are constantly subjected to such pervasive nonsense about the brain, so it is worth pausing to investigate the various sources of the 10% myth

(Nyhus and Sobel 2003). It seems to have begun with an Italian neuro-surgeon c.1890 who removed scoops of brains of psychiatric patients to see if there were any differences in their reported behaviours. The myth received an unexpected boost c.1920 during a radio interview with Albert Einstein, when the physicist used the 10% figure to implore us to think more. The myth received its widest circulation before the Second World War when some American advertisers of home-help manuals re-invented the 10% figure in order to convince customers that they were not very smart. Odd, then, that it has been so enthusiastically adopted by wishful-thinking educationists at the end of the twentieth century. It would be nice if the brains of our students had all this spare educable capacity. To be sure, the plasticity of young (and even older) brains should never be underestimated. But what plasticity requires is a dynamically engaged brain, with all neurons firing. To put it bluntly, if you are only using 10% of your brain, then you are in a vegetative state so close to death that you should hope (not that you could) that your relatives will pull out the plug of the life support machine!

Left- and right-brained thinking

Another pervasive example of over-simplification has been the misinterpretation of laterality studies to produce so-called 'left- and right-brained thinking'. Historically, the original studies were of split-brain patients: patients who had the major communication tract between the two brain hemispheres, the corpus callosum, surgically severed in an attempt to reduce life-threatening epilepsy. It was found that the separate hemispheres of these patients could separately process different types of information, but only the left hemisphere processing was reported by the patients. Unfortunately, the caveat that the researchers who carried out these studies back in the 1970s did emphasise – i.e., that these patients had abnormal brains – was largely ignored. For normal people, as Singh and O'Boyle (2004, 671) point out:

> the brain does not consist of two hemispheres operating in isolation. In fact, the different cognitive specialties of the LH and RH are so well integrated that they seldom cause significant processing conflicts ... hemispheric specialisation ... consists of a dynamic interactive partnership between the two.

Creative thinking, in particular, requires the interaction of both hemispheric specialists, neither one can operate in isolation from the other:

> Since the right hemisphere and the left hemisphere are massively interconnected (through the corpus callosum), it is not only possible, but also highly likely, that the creative person can iterate back and forth between these specialized modes to arrive at a practical solution to a real problem. If the right hemisphere were somehow disconnected from the left and confined to its own specialized thinking modes, it might be relegated to only 'soft' fantasy solutions, pipe dreams or weird ideas that would be difficult, if not impossible, to fully implement in the real world. The left brain helps keep the right brain on track. (Herrmann 1998, http://www.sciam.com)

This, then, has important implications for the misguided 'right-brain' promotion of creative thinking in the school classroom. Goswami (2004) draws attention to a recent OECD report in which left brain/right brain learning is the most troubling of several neuromyths – a sort of anti-intellectual virus which spreads among lay people as misinformation about what neuroscience can offer education. This is not to say that there isn't abundant good evidence that much brain functioning is modular, and that many higher cognitive functions, such as language production, are critically reliant on modules which are usually found in one or other hemisphere, such as Broca's Area (BA), usually found in the left frontal cortex. But there are notable differences between individuals as to where these modules are located. In about 5% of right-handed males, BA is found in the right frontal cortex, and in a higher number of females, the principle function of BA, language production, is found in both the left and right frontal cortices. In left-handed people, only 60% have BA functions on the left, with the rest having their language production involving frontal areas on both sides or on the right (Kolb and Wishaw 1990). An implication of this for neuroscience research is that practically all subjects in neuroimaging studies are screened for extreme right-handedness – it is a way of maximising the probability that the group map has contributions from all subjects (that is, their functional modules involved in the study will be in much the same place in the different individual's brains). Consequently, with a nice circularity, the data which show that language production is on the left comes almost exclusively from subjects who've been chosen to have their language production areas on the left.

Thus the left- and right-brain thinking myth seems to have arisen from misapplying lab studies which show that the semantic system is left-lateralised (language information processing in the left hemisphere; graphic and emotional information processing in the right hemisphere) by ignoring several important caveats. First, the left-lateralisation is in fact a statistically significant bias, not an absolute. Even in left-lateralised individuals, language processing does stimulate some right hemisphere activation. Second, the subjects for such studies are extremely right-handed. As language researchers are at pains to point out: 'It is dangerous to suppose that language processing only occurs in the left hemisphere of all people' (Thierry, Giraud, and Price 2003, 506). The largest interconnection to transmit information in the brain is the corpus callosum, the thick band of fibres which connects the two hemispheres. It seems that the left and right sides of our brains cannot help but pass all information between them. In fact, there is some evidence that constrictions in the corpus callosum could be predictive of deficiencies in reading abilities (Fine 2005), which obviously could not occur if language processing was an exclusively left hemisphere activity.

It would be neat if all cognitive functioning was simply lateralised, and towards such a schema some commentators have suggested that perhaps there are stylistic differences between left and right hemispheric functions, with the left mediating detail, while the holistic right focuses on the bigger picture. For example, using EEG to describe the time course of activations identified by fMRI, Jung-Beeman et al. (2004) found that the insight or 'aha' moment of problem solution elicits increased neural activity in the right hemisphere's temporal lobe. Jung-Beeman et al. (2004) suggest that the this right hemisphere function facilitates a coarse-level integration of information from distant relational sources, in contrast to the finer-level information processing characteristic of its left hemisphere homologue. However, researchers in music cognition disagree (Peretz 2003). Even regarding the left hemisphere (metaphorically if not literally) as a verbal processor, music, as non-verbal information *par excellence*, is not exclusively processed in the right, but in both hemispheres (Peretz 2003). Moreover, neuroimaging studies have shown that the location and extent of various areas of the brain involved with music perception and production shift and grow with musical experience (Parsons 2003). In fact, there is a strong evolutionary argument that music plays a crucial role in promoting the growth of the inter-module connections which underpin cognitive development in infants and young children (Cross 1999).

Consequently, for the many reasons noted above, leading neuroscientists have been calling on the neuroscience community to shift their interpretative focus of brain function from modularisation to interaction. As Hellige (2000, 206) pleads: 'Having learned so much about hemispheric differences ... it is now time to put the brain back together again.' Or as Walsh and Pascual-Leone (2003, 206) summarise: 'Human brain function and behaviour seem best explained on the basis of functional connectivity between brain structures rather than on the basis of localization of a given function to a specific brain structure.'

VAK learning styles

This emphasis on connectedness rather than separateness of brain functions has important implications for education (Geake 2004). The multi-sensory pedagogies, which experienced teachers know to be effective, are supported by fMRI research. The work of Calvert, Campbell and Brammer (2000), on imaging brain sites of cross-modal binding in human subjects, seems relevant. Bimodal processing of congruent information has a supra-additive effect (e.g., simultaneously seeing and hearing the same information works better than first just seeing and then hearing it). These findings are consistent with observed behaviour. Much good pedagogy in the early years of schooling is based on coincident bimodal information processing, especially sight and sound, or sight and speech, as demonstrated by every early years teacher pointing to the words of the story as she reads them aloud.

However, such 'natural' pedagogy is threatened by the promulgation of learning styles. The notion that individual differences in academic abilities can be partly attributed to individual learning styles has considerable intuitive appeal if we are to judge by the number of learning style models or inventories that have been devised – 170 at the last count, and rising (Coffield et al. 2004). The myriad ways that approaches to learning can seem to be partitioned, labelled and measured seems to know no bounds. The disappointing outcome of all of this endeavour is that, overall, the evidence consistently shows that modifying a teaching approach to cater for differences in learning styles does not result in any improvement in learning outcomes (Coffield et al. 2004).

Despite the lack of positive evidence, the education community has been swamped by claims for a learning style model based on the sensory modalities: visual, auditory and kinaesthetic (VAK) (Dunn, Dunn and Price 1984). The idea is that children can be tested to ascertain which is their dominant learning style, V, A or K, and then taught accordingly. Some schools have even gone so far as to label children with V, A and K shirts, presumably because these purported differences are no longer obvious in the classroom. The implicit assumption here is that the information gained through one sensory modality is processed in the brain to be learned independently from information gained through another sensory modality. There is plenty of evidence from a plethora of cross-modal investigations as to why such an assumption is wrong. What is possibly more insidious is that focusing on one sensory modality flies in the face of the brain's natural interconnectivity. VAK might, if it has any effect at all, be actually harming the academic prospects of the children so inflicted.

A simple demonstration of the ineffectiveness of VAK as a model of cognition comes from asking 5-year-olds to distinguish different sized groups of dots where the groups are too large for counting (Gilmore, McCarthy, and Spike 2007). So long as the group sizes are not almost equal, young children can do this quite reliably. Now, what happens when one group is replaced by as many sounds played too rapidly for counting? There is no change in accuracy! Going from a V versus V version of the task to a V versus A version makes no difference to task performance. The reason is that input modalities in the brain are interlinked: visual with auditory; visual with motor; motor with auditory; visual with taste; and so on. There are well-adapted evolutionary reasons for this. Out on the savannah as a pre-hominid hunter-gatherer, coordinating sight and sound makes all the difference between *detecting* dinner and *being* dinner. As Sherrington (1938, 217) noted:

> The naïve observer would have expected evolution in its course to have supplied us with more various sense organs for ampler perception of the world... Not new senses but better liaison between the old senses is what the developing nervous system has in this respect stood for.

To emphasise the cross-modal nature of sensory experience, Kayser (2007) writes that: 'the brain sees with its ears and touch, and hears with its eyes.' Moreover, as primates, we are predominantly processors of visual information. This is true even for congenitally blind children who instantiate Braille not in the kinaesthetic areas of their brains, but in those parts of their visual cortices that sighted children dedicate to learning written language. Moreover, unsighted people create the same mental spatial maps of their physical reality as sighted people do (Kriegseis et al. in press). Obviously the information to create spatial maps by blind people comes from auditory and tactile inputs, but it gets used as though it was visual. Similarly, people who after losing their hearing get a cochlear implant find that they are suddenly much more dependent on visual speech, such as cues for segmentation and formats, to conduct conversation (Thomas and Pilling in press). Wright (2007) points out just how interconnected our daily neural processes must be. Eating does not engage just taste, but smell, tactile (inside the mouth), auditory and visual sensations. Learning a language, and the practice of it, requires the coordinated use of visual, auditory and kinaesthetic modalities, in addition to memory, emotion, will, thinking and imagination:

> To an anatomist this implies the need for an immense number of neural connections between many parts of the brain. In particular, there must be numerous links between the primary auditory cortex (in the temporal lobe), the primary proprioceptive-tactile cortex (in the parietal lobe) and the primary visual cortex (in the occipital lobe). There is indeed such a neural concourse, in the parieto-temporo-occipital 'association' cortex in each cerebral hemisphere. (Wright 2007, 275)

Input information is abstracted to be processed and learnt, mostly unconsciously, through the brain's interconnectivity (Dehaene, Kerszberg, and Changeux 1998). Actually, we don't even create sensory perception in our sensory cortices:

> For a long time it was thought that the primary sensory areas are the substrate of our perception.... these zones simply generate representational maps of the sensorial information... although these respond to stimuli, they are not responsible for... perceptions... Perceptual experience occurs in certain zones of the frontal lobes [where] neurons combine sensory information with memory information. (Trujillo 2006, M9)

Literally following a VAK regime in real classrooms would lead to all sorts of ridiculous paradoxes: what does a teacher do with: the V and K 'learners' in a music lesson/ the A and K 'learners' at an art lesson/ the V and A 'learners' in a craft practical lesson? The images of blindfolds and corks in mouths are all too reminiscent of *Tommy*, the rock opera by The Who. As Sharp, Byrne and Bowker (in press) elaborate, VAK trivialises the complexity of learning, and in doing so, threatens the professionality of educators. Fortunately, many teachers have not been taken in. Ironically, VAK has become, in the hands of practitioners, a recipe for a mixed-modality pedagogy where lessons have explicit presentations of material in V, A and K modes. Teachers quickly observed that their pupils' so-called learning styles were not stable, that the expressions of

V-, A- and K-ness varied with the demands of the lessons, as they should (Geake 2006). As with other learning-style inventories, research has shown that there is no improvement of learning outcomes with VAK above teacher enthusiasm, where 'attempts to focus on learning styles were wasted effort' (Kratzig and Arbuthnott 2006).

We might speculate in passing why do VAK and other 'learning styles' seem so attractive? I wonder if two aspects of folk psychology, where we seem to learn differently from each other, and we have five senses, have created folk neuroscience: the working of our brains directly reflects our folk psychology. Of course, if our brains were that simple, we wouldn't be here today!

References

Beere, J. 2006. Capturing hearts and minds. In *Oxfordshire Governor*. Autumn: 8–9.

Beyerstein, B.L. 2004. Ask the experts: Do we really use only 10% of our brains? *Scientific American* 290, no. 6: 86.

Blakemore, S.-J., and U. Frith. 2005. *The learning brain: Lessons for education.* Oxford: Blackwell.

Calvert, G.A., R. Campbell, and M.J. Brammer. 2000. Evidence from functional magnetic resonance imaging of crossmodal binding in human heteromodal cortex. *Current Biology* 10, no. 11: 649–57.

Carroll, J. 1993. *Human cognitive abilities: A survey of factor-analytic studies.* Cambridge, UK: Cambridge University Press.

Coffield, F., D. Moseley, E. Hall, and K. Ecclestone. 2004. *Learning styles and pedagogy in post-16 learning: A systematic and critical review.* Report No. 041543. London: Learning and Skills Research Centre.

Cross, L. 1999. Is music the most important thing we ever did? Music, development and evolution. In *Music, mind and science*, ed. S.W. Yi. Seoul: Seoul National University Press.

Dehaene, S., M. Kerszberg, and J.-P. Changeux. 1998. A neuronal model of a global workspace in effortful cognitive tasks. *Proceedings of the National Academy of Sciences USA* 95: 14529–34.

Duncan, J. 2001. An adaptive coding model of neural function in prefrontal cortex. *Nature Reviews Neuroscience* 2, no. 11: 820–9.

Duncan, J., R.J. Seitz, J. Kolodny, D. Bor, H. Herzog, A. Ahmed, F.N. Newell, and H. Emslie. 2000. A neural basis for general intelligence. *Science* 289: 457–60.

Dunn, R., K. Dunn, and G.E. Price. 1984. *Learning style inventory.* Lawrence, KS: Price Systems.

Feynman, R.P. 1974. Cargo cult science. *Engineering and Science*, June: 10–3.

Fine, J. 2005. *Reading deficits predicted by differences in structure of the corpus callosum.* Montreal: AERA.

Gardner, H. 1993. *Multiple intelligences: The theory in practice.* New York: Basic Books.

Geake, J.G. 2004. How children's brains think: Not left or right but both together. *Education 3–13* 32, no. 3: 65–72.

Geake, J.G. 2006. The neurological basis of intelligence: A contrast with 'brain-based' education. *Education-Line.* http://www.leeds.ac.uk/educol/documents/156074.htm.

———. in press. Neural interconnectivity and intellectual creativity: Giftedness, savants, and learning styles. In *Companion to gifted education*, ed. T. Balchin and B. Hymer. London: Routledge.

Geake, J.G., and C.S. Dobson. 2005. A neuro-psychological model of the creative intelligence of gifted children. *Gifted and Talented International* 20, no. 1: 4–16.

Gilmore, C.K., S.E. McCarthy, and E. Spike. 2007. Symbolic arithmetic knowledge without instruction. *Nature* 447: 589–91.

Goswami, U. 2004. Neuroscience and education. *British Journal of Educational Psychology* 74: 1–14.

Hellige, J.B. 2000. All the king's horses and all the king's men: Putting the brain back together again. *Brain and Cognition* 42: 7–9.

Herrmann, N. 1998. Is it true that creativity resides in the right hemisphere of the brain? *Scientific American: Ask the Experts* 26, January. http://www.sciam.com (accessed 26 January 2008).

Howard-Jones, P. 2007. *Neuroscience and education: Issues and opportunities. Commentary by the Teacher and Learning Research Programme.* London: TLRP. http://www.tlrp.org/pub/commentaries.html.

Jung-Beeman, M., E.M. Bowden, J. Haberman, J.L. Frymiare, S. Arambel-Liu, R. Greenblatt, P.J. Reber, and J. Kounios. 2004. Neural activity when people solve verbal problems with insight. *Public Library of Science Biology* 2: 500–10.

Kayser, C. 2007. Listening with your eyes. *Scientific American Mind* 18, no. 2: 24–9.

Kratzig, G.P., and K.D. Arbuthnott. 2006. Perceptual learning style and learning proficiency: A test of the hypothesis. *Journal of Educational Psychology* 98, no. 1: 238–46.

Kolb, B., and I. Wishaw. 1990. *Fundamentals of human neuropsychology.* 3rd edn. New York: Freeman.

Kriegseis, A., E. Hennighausen, F. Rösler, and B. Röder. In press. Reduced EEG alpha activity over parieto-occipital brain areas in congenitally blind adults. *Clinical Neurophysiology.*

Mithen, S.J. 2005. *The singing Neanderthals: The origins of music, language, mind and body.* London: Weidenfeld & Nicolson.

Nyhus, E.M., and N. Sobel. 2003. 'The 10% Myth'. Poster presented at the Society for Neuroscience Conference, New Orleans.

Parsons, L.M. 2003. Exploring the functional neuroanatomy of music performance, perception, and comprehension. In *The cognitive neuroscience of music,* ed. I. Peretz and R. Zatorre. Oxford: Oxford University Press.

Peretz, I. 2003. Brain specialisations for music: New evidence from congenital amusia. In *The cognitive neuroscience of music,* ed. I. Peretz and R. Zatorre. Oxford: Oxford University Press.

Pickering, S.J., and P.A. Howard-Jones. 2007. Educators' views of the role of neuroscience, in education: A study of UK and international perspectives. *Mind, Brain and Education* 1, no. 3: 110–13.

Sharp, J.G., J. Byrne, and R. Bowker. In press. VAK or VAK-uous? Lessons in the trivialisation of learning and the death of scholarship. *Research Papers in Education.*

Sherrington, C. 1938. *Man on his nature.* Cambridge, UK: Cambridge University Press.

Singh, H., and M.W. O'Boyle. 2004. Interhemispheric interaction during global–local processing in mathematically gifted adolescents, average-ability youth, and college students. *Neuropsychology* 18, no. 2: 671–7.

Thierry, G., A.L. Giraud, and C. Price. 2003. Hemispheric dissociation in accessing the human semantic system. *Neuron* 38: 499–506.

Thomas, S.M., and M. Pilling. In press. Auditory and audiovisual training for cochlear-implant simulation. *Ear and Hearing.*

Trujillo, R.R. 2006. Discovering the brain's language. *Scientific American* 295, no. 3: M9.

Walsh, V., and A. Pascual-Leone. 2003. *Transcranial magnetic stimulation: A neurochronometrics of mind.* London: The MIT Press.

Waterhouse, L. 2006. Multiple intelligences, the Mozart effect, and emotional intelligence: A critical review. *Educational Psychologist* 41, no. 4: 207–25.

Wright, G. 2007. *The anatomy of metaphor.* Cambridge, UK: Clare College.

Reading, dyslexia and the brain

Usha Goswami

Centre for Neuroscience in Education, University of Cambridge, UK

Background: Neuroimaging offers unique opportunities for understanding the acquisition of reading by children and for unravelling the mystery of developmental dyslexia. Here, I provide a selective overview of recent neuroimaging studies, drawing out implications for education and the teaching of reading.

Purpose: The different neuroimaging technologies available offer complementary techniques for revealing the biological basis of reading and dyslexia. Functional magnetic resonance imaging (fMRI) is most suited to localisation of function, and hence to investigating the neural networks that underpin efficient (or inefficient) reading. Electroencephalography (EEG) is sensitive to millisecond differences in timing, hence it is suited to studying the time course of processing; for example, it can reveal when networks relevant to phonology versus semantics are activated. Magnetic source imaging (MSI) gives information about both location in the brain and the time course of activation. I illustrate how each technology is most suited to answering particular questions about the core neural systems for reading, and how these systems interact, and what might go wrong in the dyslexic brain.

Design and methods: Following a brief overview of behavioural studies of reading acquisition in different languages, selected neuroimaging studies of typical development are discussed and analysed. Those studies including the widest age ranges of children have been selected. Neuroimaging studies of developmental dyslexia are then reviewed, focusing on (a) the neural networks recruited for reading, (b) the time course of neural activation and (c) the neural effects of remediation. Representative studies using the different methodologies are selected. It is shown that studies converge in showing that the dyslexic brain is characterised by *under-activation* of the key neural networks for reading.

Conclusions: Different neuroimaging methods can contribute different kinds of data relevant to key questions in education. The most informative studies with respect to causation will be longitudinal prospective studies, which are currently rare.

How universal are the neural demands made by learning to read in different languages? What are the core neural systems involved, and what goes wrong in the dyslexic brain? Current neuroimaging technologies are able to throw light on research questions such as these, as will be illustrated below. In some instances, neuroimaging technologies can contribute unique information that behavioural methodologies are simply unable to provide. This includes information about the time processes in reading, and information about the parts of the brain that are affected by remedial packages for developmental dyslexia. Some neuroimaging methodologies can gather data without requiring overt attention on the part of the child. These methodologies are particularly powerful for contributing to our understanding of the biological basis of developmental dyslexia.

The history of research on developmental dyslexia has been dominated by visual theories of the disorder, ever since Hinshelwood (1896) described it as 'congenital word blindness'. Historically, theories of reading development also assumed that visual processing was core to individual differences in the acquisition of reading. In the 1970s, for example, there was much discussion of 'Phonecian' versus 'Chinese' reading acquisition strategies. It was assumed that children who were learning to read character-based orthographies like Chinese required excellent visual memory skills in order to distinguish between the visually complex characters that represented spoken words. Hence visual memory or 'logographic' strategies were assumed core to reading acquisition of languages like Chinese and Japanese. Children who were learning to read languages like Greek or Italian, which were alphabetic and transparent (each letter corresponding to one, and only one, sound) appeared to require code-breaking skills. It was assumed that once the brain had learned the symbol–sound code, reading should be largely a process of phonological assembly. Many experiments were conducted with children learning to read in English, to compare the contribution of 'Chinese' versus 'Phonecian' acquisition strategies (e.g., Baron 1979). Dual-route models of reading, originally developed using data from adults, were applied to children who were learning to read (Stuart and Coltheart 1988). It was assumed that, developmentally, children could choose to learn to read by either Chinese or Phonecian strategies.

These ideas about individual differences have not gone away (e.g., Stein and Walsh 1997; Stuart 2006), but they are looking increasingly dated with the advent of brain imaging. Neuroimaging has also shed light on the processes underpinning the development of reading in deaf children, whom it was once assumed had no choice but to rely on visual memorisation strategies (e.g., Conrad 1979). Essentially, neuroscience is showing that despite the apparently different demands on the brain made by learning to read English, Greek or Chinese, and the apparently different processing strategies used by children who are deaf or who are dyslexic, reading across orthographies depends on the adequate functioning of the phonological system. Even for languages like Chinese, which would appear reliant on visual processing, it is oral language skills that underpin the acquisition of reading.

As I show in this review, brain imaging studies demonstrate that reading begins primarily as a phonological process. In the early phases of learning to read, it is the neural structures for spoken language that are particularly active. As reading expertise develops, an area in the visual cortex originally named the 'visual word form area' (VWFA) becomes increasingly active (Cohen and Dehaene 2004). This area is not a logographic system, even though it is very close to the visual areas that are active during picture naming. The VWFA is also active during nonsense word reading, and as nonsense words do not have word forms in the mental lexicon, the VWFA is thought to store orthography–phonology connections at different grain sizes (Goswami and Ziegler 2006a). Deaf readers rely on the same phonological system for reading as everyone else (MacSweeney et al. 2005). Children with developmental dyslexia show selective underactivation of key phonological areas of the brain, but targeted phonology-based interventions improve levels of activation in these areas, 'normalising' neural activity (Simos et al. 2002).

Learning to read: behavioural data

Many behavioural studies in developmental psychology show the critical role of 'phonological awareness' in learning to read (for a recent review see Ziegler and Goswami 2005). Phonological awareness is thought to develop via language acquisition. Between the ages of 1 and 6 years, children acquire words at an exponential rate. For example, the average 1-year-old might have a productive vocabulary of around 50–100 words, but by the age of 6 the average child's receptive vocabulary will contain around 14,000 words (Dollaghan 1994). In order for the brain to represent each word as a distinct and unique sequence of sounds, each entry in the 'mental lexicon' must incorporate phonological information along with information about meaning. For example, there must be implicit knowledge of the sounds that comprise a particular word, and the order in which they occur. Phonological awareness is essentially the child's ability to reflect on this implicit knowledge, and to make judgements based on it. Hence phonological awareness is typically measured by a child's ability to detect and manipulate component sounds in words, for example, by deciding whether words rhyme, or by removing the initial sound from a spoken word.

The syllable is the primary processing unit across the world's languages (Port 2006). In fact, there is an apparently language-universal sequence in the development of phonological awareness, from syllable awareness, through 'onset-rime' awareness to 'phoneme' awareness. Syllables ('university' has five syllables, 'coffee' has two syllables) can be segmented into sub-parts called onsets and rimes. The onset is the sound or sounds before the vowel, such as the 'spr' sound in 'spring' and the 'st' sound in 'sting'. The rime is the vowel and any subsequent sounds in the syllable, such as the 'ing' sound in 'spring' and 'string'. The phoneme is the smallest unit of sound that changes meaning. 'Spring' and 'string' differ in meaning because the second sound is different in each word ('p' versus 't' respectively). In many of the world's languages, onsets and rimes are the same as phonemes. This is because the dominant syllable structure across the world's languages is consonant–vowel (CV). Relatively few words in English are CV syllables (5% of English monosyllables follow a CV structure: see De Cara and Goswami 2002). Examples of English words comprised of CV syllables are 'go', 'do' and 'yoyo'.

Behavioural studies across languages have shown that phonological sensitivity at all three linguistic levels (syllable, onset-rime, phoneme) predicts the acquisition of reading (for a review see Ziegler and Goswami 2005). Furthermore, it has been demonstrated that training phonological awareness has positive effects on reading acquisition across languages, particularly when it is combined with training about how letters or letter sequences correspond to sounds in that language (e.g., Bradley and Bryant 1983; Schneider, Roth, and Ennemoser 2000). Children with developmental dyslexia across languages appear to have specific problems in detecting and manipulating component sounds in words (called a 'phonological deficit': see, e.g., Snowling 2000). For example, they find it difficult to count the number of syllables in different words, to recognise rhymes, to distinguish shared phonemes and to delete phonemes or substitute one phoneme for another (Korean: Kim and Davis 2004; German: Wimmer 1996; Greek: Porpodas 1999; Hebrew: Share and Levin 1999; for a comprehensive review see Ziegler and Goswami 2005). Dyslexic children are developing some awareness of phonology, but this is a slow and effortful process. Deaf children also develop phonological codes, for example, via lip reading ('speech reading') and vibrational cues. This is the case even if signing is their native language. Phonology is essentially the smallest contrastive units of a language that create new meanings. In signed languages, phonology depends on visual/manual elements, with handshapes, movements and locations combined to form signs (Sandler and Lillo-Martin 2006). For deaf children too, individual differences in phonological awareness are related to reading acquisition (e.g., Harris and Beech 1998).

Reading and learning to read: neuroimaging data

To date, most neuroimaging studies of reading have been conducted with adults (see Price and McCrory 2005 for a recent synthesis). This was partly due to the methodologies available. The most popular methods for studying the brain during the act of reading depended on imaging techniques like functional magnetic resonance imaging (fMRI) and positron emission tomography (PET). The fMRI technique measures changes in blood flow in the brain, which take approximately 6–8 seconds to reach a maximum value (i.e. maximum activity will be measurable 6–8 seconds after reading a particular word). fMRI works by measuring the magnetic resonance signal generated by the protons of water molecules in brain cells, generating a BOLD (blood oxygenation level dependent) response. The fMRI method is excellent for the localisation of function, but because changes in brain activity are summated over time, it cannot provide information about the sequence in which different neural networks become engaged during the act of reading. In PET, radioactive tracers are injected into the bloodstream and provide an index of brain metabolism. Because of the use of radioactive tracers, PET is not suitable for studying children.

More recently, the value of the electroencephalogram (EEG) methodology for studying reading is being recognised. Neurons communicate on a millisecond scale, with the earliest stages of cognitive information processing beginning between 100 ms and 200 ms after stimulus presentation. EEG methods can measure the low-voltage changes caused by the electro-chemical activity of brain cells, thereby reflecting the direct electrical activity of neurons at the time of stimulation (e.g., at the time of seeing a word). Initially, EEG methods were less widely used in the neuroscience of reading, because it is difficult to localise function using EEG. However, developmentally, information about the time course of processing is very important. Data from EEG studies suggest that the brain has decided whether it is reading a real word or a nonsense word within 160–180 ms of presentation, for children and adults across languages (Csepe and Szucs 2003; Suaseng, Bergmann, and Wimmer 2004).

Adult studies of reading based on PET and fMRI have focused on a relatively small range of reading and reading-related tasks, and studies of children using fMRI have followed suit. Typical tasks include asking participants to read single words and then comparing brain activation to a resting condition with the eyes closed; asking participants to pick out target visual features while reading print or 'false font' (false font is made up of meaningless symbols matched to letters for visual features like the 'ascenders' in the letters b, d, k); making phonological judgements while reading words or nonsense words (e.g., 'do these items rhyme?': leat, jete) and making lexical decisions (e.g., pressing a button when a word is presented, and a different button when a nonsense word is presented). Adult experiments show a very consistent picture concerning the neural networks that underpin skilled reading (e.g., Price et al. 2003; Rumsey et al. 1997; see for comments on divergence Price and McCrory 2005). Word recognition in skilled readers appears to depend on a left-lateralised network of frontal, temporoparietal and occipitotemporal regions, whatever language they are reading (see Figure 1). However,

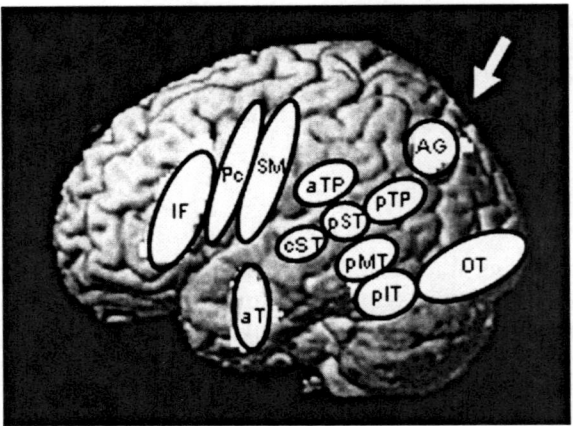

AG   = Angular Gyrus
OT   = Occipito-Temporal
pTP = posterior Temporo-Parietal
pST = posterior Superior Temporal
pIT = posterior Inferior Temporal
pMT = posterior Middle Temporal
aTP = anterior Temporo-Parietal
aT   = anterior Temporal
cST = central Superior Temporal
IF   = Inferior Frontal
Pc   = Precentral gyrus
SM   = Sensori-motor cortex

Figure 1.   A schematic depiction of some of the neural areas involved in reading (left hemisphere depiction) (from Price and McCrory 2005).

there is some additional recruitment of visuo-spatial areas for languages with non-alphabetic orthographies (e.g., left middle frontal gyrus for Chinese: see meta-analysis by Tan et al. 2005). The frontal, temporoparietal and occipitotemporal regions essentially comprise the language, auditory, cross-modal and visual areas of the brain. At a very simple level, semantic and memory processing is thought to occur in temporal and frontal areas, auditory and phonological processing in temporal areas, articulation in frontal areas, visual processing in occipital areas and cross-modal processing in parietal areas.

Although there are still relatively few neuroimaging studies of children reading, the studies that have been done show a high degree of consistency in the neural networks recruited by novice and expert readers. For example, work by Turkeltaub and colleagues has used fMRI and the false font task to compare neural activation in English-speaking participants aged from 7 to 22 years (Turkeltaub et al. 2003). Importantly, 7-year-olds can perform the 'false font' task as well as adults, hence changes in reading-related neural activity are likely to reflect developmental differences rather than differences in reading expertise. Turkeltaub et al. (2003) reported that adults performing their task activated the usual left hemisphere sites, including left posterior temporal and left inferior frontal cortex. They then restricted the analyses to children below 9 years of age. Now the main area engaged was left posterior superior temporal cortex. This region is traditionally considered the focus of phonological activity, and is thus thought to be active during grapheme–phoneme translation. As reading developed, activity in left temporal and frontal areas increased, while activity previously observed in right posterior areas declined. This pattern was interpreted as showing that reading-related activity in the brain becomes more left-lateralised with development.

In further analyses focusing just on the younger children, the researchers investigated the relationships between three core phonological skills and word processing. The three core phonological skills are usually taken to be phonological awareness, phonological memory and rapid automatised naming (RAN). I will focus on the phonological awareness findings here. Turkeltaub et al. (2003) calculated partial correlations between activated brain regions and each of these three measures while controlling for the effects of the other two measures. They reported that the three different measures correlated with three distinct patterns of brain activity. Brain activity during phonological awareness tasks appeared to depend on a network of areas in left posterior superior temporal cortex (phonology and grapheme–phoneme translation) and inferior frontal gyrus (articulation). The level of the children's phonological skills modulated the amount of activity in this network. As noted earlier, the left posterior temporal sulcus was the primary area recruited by young children at the beginning of reading development. Therefore, neuroimaging data suggest that phonological recoding to sound rather than logographic recognition is the key early reading strategy. Activity in the inferior frontal gyrus increased with reading ability. This area is also a key phonological area (Broca's area), important for the motor production of speech. Left inferior frontal gyrus is also activated when deaf children perform phonological awareness tasks silently in fMRI studies (MacSweeney et al. 2005).

An fMRI study of 119 typically developing readers aged from 7 years to 17 years by Shaywitz and colleagues found a similar developmental pattern (Shaywitz et al. 2007). Instead of the false font task, this study used a rhyme decision task (e.g., 'do these items rhyme?': leat, kete), and a visual line orientation task (e.g., 'Do [\\V] and [\\V] match?'). Shaywitz and his colleagues reported that networks in both left and right superior and middle frontal regions were more active in younger readers, with activity declining as reading developed. In contrast, activity in the left anterior lateral occipitotemporal region increased. This region includes the putative visual word form area (VWFA). Hence both Turkeltaub et al. (2003) and Shaywitz et al. (2007) found decreased right hemisphere involvement as reading developed, but found this for somewhat different neural networks. The difference in the behavioural tasks used (e.g., false font versus rhyme judgement) may explain some of these differences.

Overall, therefore, current neuroimaging data support a 'single route' model of reading development, based on a process of developing orthographic–phonological connections at different grain sizes (Ziegler and Goswami 2006). Reading is founded in phonology from the beginning (Goswami and Ziegler 2006b). The VWFA becomes more active as reading develops, reflecting the development of an orthographic lexicon containing both whole words and fragments of familiar words such as orthographic rimes (Pugh 2006). The VWFA is not a logographic or visual lexicon, able to support 'Chinese' processing or the 'direct route' from printed word to meaning postulated by 'dual-route' theory. Neuroimaging studies of typically developing readers show that the neural networks for spoken language play an important developmental role in reading from the outset.

Neuroimaging studies of dyslexia

The networks recruited for reading

Neuroimaging studies of adult readers with developmental dyslexia suggest that there is atypical activation in the three important neural sites for reading, namely the left posterior temporal regions, the left inferior frontal regions and the left occipitotemporal regions (such as the VWFA). These data suggest both problems with the phonological aspects of reading and with the efficient development of an orthographic lexicon (e.g., Brunswick et al. 1999). These fMRI and PET studies typically rely on tasks such as word and nonsense word reading (e.g., 'valley', 'carrot', 'vassey', 'cassot'), and the 'false font' task. Again, the experimental picture is largely one of convergence across orthographies. For example, adult dyslexics in Italian, French and English all showed activation of a left-lateralised neural network based around posterior inferior temporal areas and middle occipital gyrus (Paulesu et al. 2001). This was a cross-language comparison within one study. However, issues of experimental design become critical when comparing individual imaging studies across languages. When studying any kind of disability, it is crucial to equate participant groups for their overall ability in the actual tasks being used to acquire the neuroimaging data. For example, it is impossible to interpret group differences in brain activity if the dyslexics are worse at reading the nonsense words being used than the control adults. In this case, differences in neural activation could simply reflect different skill levels (i.e., behavioural differences in reading performance). Similarly, it is critical to use the same criteria for acquiring images of the brain in different studies if interpretations about cross-language differences are being drawn (e.g., Ziegler 2005). Otherwise, apparent language-based differences could simply reflect differences in the significance thresholds or other experimental criteria used to acquire the images by different research groups.

Neuroimaging studies of children with developmental dyslexia report a very similar pattern to adult data (e.g., Shaywitz et al. 2002, 2007; Simos et al. 2000). For example, Shaywitz et al. (2002) studied 70 children with dyslexia aged on average 13 years, and compared them to 74 11-year-old typically developing controls (although the controls were not matched for reading level). Using fMRI, the children were scanned while performing a variety of reading-related tasks. These were letter identification (e.g., are t and V the same letter?); single letter rhyme (e.g., do V and C rhyme?); non-word rhyming (e.g., do leat and jete rhyme?); and reading for meaning (e.g., are corn and rice in the same semantic category?). Brain activity in each condition was contrasted with activity in a baseline condition, the line orientation task (e.g., do [\\V] and [\\V] match?). Shaywitz et al. (2002) reported that the children with developmental dyslexia showed under-activation in the core left temporoparietal networks, with older dyslexics showing over-activation in right inferior frontal gyrus. The children with developmental dyslexia also showed increased activation in right temporoparietal networks. One drawback of the study, however, was that there were group differences in behavioural performance in some of the component tasks. In the non-word rhyming measure, for example, the controls [79%] were significantly better than the children with dyslexia [59%]). This means that some of the differences found in brain activation could reflect differing levels of expertise rather than differences core to having developmental dyslexia. In a subsequent study of an expanded sample, Shaywitz et al. (2007) used in-magnet non-word reading ability as a covariate to control for this problem. Shaywitz et al. compared 113 dyslexic children aged 7–18 years to the 119 typically developing readers discussed above in the non-word rhyme and visual line orientation tasks. Compared to the typically developing children, the dyslexic children showed no age-related increase in the activity of the VWFA. Instead, activity in the left inferior frontal gyrus (speech articulation) and the left posterior medial occipitotemporal system both increased, and reading did not become left-lateralized, with continued right hemisphere involvement.

There are also a few studies in the literature exploring the neural networks recruited for reading by dyslexic children in other languages. A study of 13 German dyslexic children aged 14–16 years was reported by Kronbichler et al. (2006). They used a sentence verification task (e.g., 'A flower needs water' – TRUE), in an fMRI design, to try and replicate natural reading. A false font task provided the control task. Consistent with studies of English dyslexics, they found reduced activation of left occipitotemporal networks and increased activation of left inferior frontal areas. A study of eight Chinese children with developmental dyslexia reported by Siok et al. (2004) claimed biological disunity, however. Their fMRI study used a homophone judgement task, in which the children had to decide whether two different Chinese characters made the same sound (an English homophone is <u>week</u> – <u>weak</u>), and a character decision task, in which the children saw one Chinese character and had to decide whether it was a real word or not. The first task was intended to measure orthography–phonology connections, and the second orthography–semantic relations. Siok et al. (2004) reported that the Chinese dyslexics did not demonstrate the reduced activation in left temporoparietal regions that would typically be found in developmental dyslexia in English during the homophone judgement task. Instead, an area involved in visuo-spatial analysis showed reduced activity, the left middle frontal gyrus. Siok et al. (2004) used this latter finding to argue that the biological marker for developmental dyslexia in Chinese was reduced activation of left middle frontal gyrus. However, the design of this study does not yet permit this conclusion. A control group matched for reading level is also required. Reduced activation in left middle frontal gyrus when making homophone judgements in Chinese might be expected for the level of reading achieved by the children with dyslexia. If this were to be the case, then increased involvement of networks for visuo-spatial analysis as reading develops would be part of typical reading development in Chinese, rather than a unique biological marker for developmental dyslexia.

Developmental differences in the time course of neural activation

While fMRI studies can provide important information about the neural networks supporting reading in typically developing versus dyslexic readers, they do not provide information about the time course of neural processing. This is important, as in typically developing readers words are distinguished from non-words within around 180 ms, suggesting early contact with the VWFA and semantic sites. It seems likely that this process would be delayed in developmental dyslexia. Similarly, it seems possible that cognitive processes such as grapheme–phoneme conversion might take longer in developmental dyslexia.

A longitudinal study of 33 English-speaking children using magnetic source imaging (MSI) compared brain activation in a letter–sound task (the child sees a letter and has to provide its sound) and a simple non-word reading task (e.g., 'lan') at the end of kindergarten and again at the end of grade 1 (Simos et al. 2005). Magnetic source imaging depends on a combination of magneto-encephalography (MEG) and MRI. The MEG measures the magnetic fields generated by the electrical activity in the brain rather than the electrical activity itself (the latter is measured by EEG). These magnetic fields are tiny, they are one billion times smaller than the magnetic field generated by the electricity in a lightbulb. By combining this information with MRI scans, both the time course and spatial localisation of brain activity is possible. Of the 33 children studied, 16 were thought to be at high risk of developing dyslexia.

Simos et al. (2005) reported that the high-risk group were significantly slower to show neural activity in response to both letters and non-words in kindergarten in the occipitotemporal region (320 ms compared to 210 ms for those not at risk). The high-risk group also showed atypical activation in the left inferior frontal gyrus when performing the letter–sound task, with the onset of activity *increasing* from 603 ms in kindergarten to 786 ms in grade 1. The typically developing readers did not show this processing time increase. Comparing the onset of activity of the three core neural networks for reading, Simos et al. (2005) reported that low-risk children showed early activity in the left occipitotemporal regions, followed by activity in temporoparietal regions, predominantly in the left hemisphere, and then bilateral activity in inferior frontal regions. In contrast, high-risk children showed little differentiation in terms of the time course of activation between the occipitotemporal and temporoparietal regions. High-risk children who were non-responsive to a phonological remediation package also being administered ($n = 3$) were distinct in showing earlier onset of activity in inferior frontal gyrus compared to the temporoparietal regions. Given the current dearth of time-course studies by other research groups in either English or in other languages, it is difficult to interpret these differences in terms of the cognitive components of reading. Nevertheless, Simos et al. (2005) comment that the increased inferior frontal activation probably reflects the role of compensatory articulatory processes. As noted earlier, deaf children also show increased inferior frontal activation during phonological processing tasks. This may indicate that children with phonological difficulties rely more heavily on networks for articulation when phonological processing is required.

The neural effects of remediation

Although there are a variety of remediation packages for dyslexic children based on different theories of developmental dyslexia, the most effective packages across languages appear to be those offering intensive phonological intervention (e.g., Bradley and Bryant 1983; Schneider, Roth, and Ennemoser 2000). Simos and his research group (2002) used magnetic source imaging to explore neural activation in eight children with developmental dyslexia who had received 80 hours of intensive training with such a package and who had shown significant benefits from the remediation (Simos et al. 2002). MSI scans were taken during a non-word rhyme matching task (e.g., 'yoat', 'wote') both before the intervention and following remediation. Simos et al. (2002) reported that prior to the intervention, the dyslexic children showed the expected hypoactivation of left temporoparietal regions. Following the intervention, all eight children showed a dramatic increase in the activation of left temporoparietal regions, predominantly in the left posterior superior temporal gyrus (the networks supporting grapheme–phoneme recoding in typically developing readers: see Turkeltaub et al. 2003). These activation profiles were very similar to those of eight controls who also participated in the MSI study, but who did not require remediation. Nevertheless, even after remediation neural activity was delayed in the children with dyslexia relative to the controls. The peak in left superior temporal gyrus activity occurred at 837 ms on average for the dyslexic children, and at 600 ms for the controls. The data were taken to show a normalisation of brain function with remediation. Nevertheless, Simos et al. (2002) commented that even with intensive remediation, children with dyslexia are slow to achieve the reading fluency shown by non-dyslexic children.

Shaywitz and Shaywitz (2005) used retrospective examination of the large sample of children with developmental dyslexia reported in Shaywitz et al. (2002) to compare the different developmental trajectories for children at risk for reading difficulties. Shaywitz and Shaywitz (2005) distinguished three groups within this sample when they were young adults. The first was a group of persistently poor readers (PPR), who had met criteria for poor reading in both the 2nd/3rd and the 9th/10th grades. The second was a group of accuracy-improved poor readers (AIR), who had met criteria for poor reading in the 2nd/3rd grades but who did not meet criteria in the 9th/10th grades. The third was a control group of non-impaired readers (C), who had never met criteria for poor reading (the participants had been studied since the age of 5 years). Shaywitz and Shaywitz (2005) reported that both the PPR and the AIR groups showed hypoactivation of the core left hemisphere sites when required to manipulate phonology. For example, in a nonsense word rhyming task, both groups of young adults still showed relative hypoactivity in neural networks in left superior temporal and occipitotemporal regions. However, the groups were distinguished by their neural activity when reading real words. The AIR group still demonstrated under-activation in the usual left posterior areas for real word reading, whereas the PPR group activated the left posterior regions to the same extent as controls (this was an unexpected finding).

Shaywitz and Shaywitz (2005) then carried out further analyses based on connectivity. Connectivity analyses examine the neural areas that are functionally connected to each other during reading. The connectivity analyses suggested that reading achievement depended on memory for the PPR group, and not on the normalised functioning of the left posterior regions. The unimpaired controls demonstrated functional connectivity between left hemisphere posterior and anterior reading systems, but the PPR group demonstrated functional connectivity between left hemisphere posterior regions and right prefrontal areas associated with working memory and memory retrieval. Shaywitz and Shaywitz (2005) speculated that the PPR group were reading primarily by memory. As the words used in the scanner were high-frequency, simple words, this is quite possible. However, this design choice complicates the interpretation of the neural differences found, as the PPR group may not be able to use memory strategies to read less frequent or less simple words. For such stimuli, the PPR and AIR groups may show similar neural profiles. It may also be important that the PPR group had, in general, lower IQ scores than the AIR group. Prospective longitudinal studies comparing patterns of neural activation and connectivity in dyslexic children as high-frequency words become over-learned would clearly be very valuable.

Different technologies, different research questions: the promise of brain imaging for understanding reading and developmental dyslexia

As will be clear from the foregoing review, most studies of reading development and of developmental dyslexia have relied on fMRI. These studies have provided excellent data regarding the neural networks underpinning reading in typically developing and dyslexic readers. They have also shown that the functional organisation of the networks for reading is similar in typical development and in dyslexia. Children with developmental dyslexia do not recruit radically different neural networks when they are reading. Rather, they show hypoactivation of crucial parts of the network of areas involved in word recognition, and an atypical pattern of continuing right hemisphere involvement. Although highly informative, these studies are essentially correlational studies. They can answer research questions about the neural demands made by learning to read in different languages, and they can answer research questions about the core neural systems involved for dyslexic and typically developing readers. They can also answer research questions about the patterns of connectivity between different neural networks. However, they cannot answer research questions about what 'goes wrong' in the dyslexic brain, although they can help to rule out hypotheses (e.g., about the visual basis of developmental dyslexia; see Eden and Zeffiro 1998).

Neuroimaging methods that provide data on the time course of neural processing, such as MEG (MSI) and EEG, can begin to answer causal questions. As might be expected, it has been shown using MSI that neural activation is delayed in core components of the network of areas recruited for reading by children at risk for dyslexia. However, behavioural studies showing that children with developmental dyslexia are slower to read words aloud make the same point. When EEG or MSI techniques show that core components of the reading network are activated in a different order in dyslexia compared to typical reading, this is more informative with respect to causality. For example, Simos and his colleagues have shown atypically earlier onset of activity in inferior frontal gyrus (articulation) compared to the temporoparietal regions in three children at high risk for dyslexia who appear to be non-respondent to a phonological remediation package. If robust with larger samples and diagnosed dyslexics, such findings could suggest that there are different neuro-developmental routes to word recognition for dyslexic children compared to controls. Nevertheless, these different neuro-developmental routes are not the cause of dyslexia. Rather, they illustrate the response of a dyslexic brain to being trained to learn to read.

In my view, the most informative studies with respect to causation in developmental dyslexia are longitudinal prospective studies that use brain imaging to study basic sensory processing in at-risk children, with a view to understanding the causes of the phonological deficit. Here, the most promising studies to date are those investigating basic auditory processing using methodologies sensitive to the time course of auditory processing at the millisecond level. For example, a large-scale Finnish study (the Jyväskylä Longitudinal Study of Dyslexia (JLD): see Lyytinen et al. 2004a) has followed babies at familial risk for dyslexia since birth. A large variety of behavioural and EEG measures has been taken as the children have developed. EEG measures of auditory sensory processing (evoked response potentials to speech and non-speech cues) have been found to distinguish the at-risk babies from controls even during infancy (e.g., Lyytinen et al. 2005). For example, infants at risk for developmental dyslexia were less sensitive to the auditory cue of duration at six months of age (Richardson et al. 2003). The infant participants had to discriminate between two bisyllabic speech-like stimuli with a varying silent interval (e.g. 'ata' versus 'atta'). Duration discrimination was still impaired when the same children were 6.5 years of age (Lyytinen et al. 2004b).

English children with developmental dyslexia are also impaired in this duration discrimination task (Richardson et al. 2004). In addition, English children are impaired in discriminating the rise time of amplitude envelopes at onset, which is an important auditory cue to the onset of syllables in the speech stream (Goswami et al. 2002; Richardson et al. 2004). Finnish adults with developmental dyslexia also show rise time processing impairments, and individual differences in rise time sensitivity predicted up to 35% of unique variance in phonological tasks like rhyme recognition (Hämäläinen et al. 2005). In the English studies, individual differences in rise time sensitivity predict unique variance in both phonological awareness measures (around 20%: Richardson et al. 2004) and in reading and spelling measures (around 25%: Goswami et al. 2002). We are currently collecting EEG data comparing rise time discrimination in English children with and without dyslexia. Data so far suggest that children with developmental dyslexia indeed show atypical auditory processing of rise time stimuli, with N1 amplitude (an EEG measure of sound registration) failing to reduce as amplitude envelope rise times become extended (Thomson, Baldeweg, and Goswami 2005). This suggests that neural responses in the dyslexic brain do not distinguish between different rise times, at least for the auditory processing comparisons used in our study (15 ms versus 90 ms rise times).

Conclusion

Different neuroimaging methodologies contribute complementary data regarding the neural networks underpinning reading acquisition and developmental dyslexia. While fMRI studies can identify the core neural systems involved in reading, EEG and MEG methodologies are required to investigate the time course of activation of the different networks that contribute to word recognition, and to investigate potential sensory precursors of the phonological deficit. With respect to key questions in education, each neuroimaging method can contribute different kinds of data. For example, when evaluating the claims made for different kinds of remediation package for developmental dyslexia, fMRI will be useful in assessing whether interventions affect the core neural networks for reading, or affect a different kind of network (e.g., motivational systems). When evaluating claims that the core cognitive difficulty in developmental dyslexia lies in forming a high-quality phonological representation, methodologies that can explore the time course of sensory processing such as EEG will be most useful. Neuroimaging methods are of optimal use when they can provide experimental data that is not available from behavioural investigations. For example, it is possible in principle to identify neural markers of risk for developmental dyslexia that can be measured in pre-verbal infants and in older children without requiring their explicit attention (Szücs and Goswami 2007). It is these areas of neuroscience that are likely to be of most potential benefit to educators.

References

Baron, J. 1979. Orthographic and word-specific mechanisms in children's reading of words. *Child Development* 50: 60–72.

Bradley, L., and P.E. Bryant. 1983. Categorising sounds and learning to read: A causal connection. *Nature* 310: 419–21.

Brunswick, N., E. McCrory, C.J. Price, C.D. Frith, and U. Frith. 1999. Explicit and implicit processing of words and pseudowords by adult developmental dyslexics: A search for Wernicke's Wortschatz. *Brain* 122: 1901–17.

Cohen, L., and S. Dehaene. 2004. Specialization within the ventral stream: The case for the visual word form area. *NeuroImage* 22: 466–76.

Conrad, R. 1979. *The deaf school child*. London: Harper & Row.

Csepe, V., and D. Szucs. 2003. Number word reading as a challenging task in dyslexia? An ERP study. *International Journal of Psychophysiology* 51: 69–83.

De Cara, B., and U. Goswami. 2002. Statistical analysis of similarity relations among spoken words: Evidence for the special status of rimes in English. *Behavioural Research Methods and Instrumentation* 34, no. 3: 416–23.

Dollaghan, C.A. 1994. Children's phonological neighbourhoods: Half empty or half full? *Journal of Child Language* 21: 257–71.

Eden, G.F., and T.A. Zeffiro. 1998. Neural systems affected in developmental dyslexia revealed by functional neuroimaging. *Neuron* 21: 279–82.

Goswami, U., J. Thomson, U. Richardson, R. Stainthorp, D. Hughes, S. Rosen et al. 2002. Amplitude envelope onsets and developmental dyslexia: A new hypothesis. *Proceedings of the National Academy of Sciences of the United States of America* 99: 10911–16.

Goswami, U., and J.C. Ziegler. 2006a. Fluency, phonology and morphology: A response to the commentaries on becoming literate in different languages. *Developmental Science* 9, no. 5: 451–3.

————. 2006b. A developmental perspective on the neural code for written words. *Trends in Cognitive Sciences* 10, no. 4: 142–3.

Hämäläinen, J., P.H.T. Leppanen, M. Torppa, K. Muller, and H. Lyytinen. 2005. Detection of sound rise time by adults with dyslexia. *Brain and Language* 94: 32–42.

Harris, M., and J.R. Beech. 1998. Implicit phonological awareness and early reading development in prelingually deaf children. *Journal of Deaf Studies and Deaf Education* 3, no. 3: 205–16.

Hinshelwood, J.A. 1896. A case of dyslexia: A peculiar form of word-blindness. *Lancet* 2: 1451.

Kim, J., and C. Davis. 2004. Characteristics of poor readers of Korean Hangul: Auditory, visual and phonological processing. *Reading and Writing* 17, no. 1–2: 153–85.

Kronbichler, M., et al. 2006. Evidence for a dysfunction of left posterior reading areas in German dyslexic readers. *Neuropsychologia* 44: 1822–32.

Lyytinen, H., T. Ahonen, and T. Guttorm, et al. 2004a. Early development of children at familial risk for dyslexia: Follow-up from birth to school age. *Dyslexia* 10: 146–78.

Lyytinen, H., M. Aro, and K. Eklund, et al. 2004b. The development of children at familial risk for dyslexia: Birth to school age. *Annals of Dyslexia* 5, no. 4: 185–220.

Lyytinen, H., et al. 2005. Psychophysiology of developmental dyslexia: A review of findings including studies of children at risk for dyslexia. *Journal of Neurolinguistics* 18: 167–95.

MacSweeney, M., D. Waters, M. Brammer, B. Woll, and U. Goswami. 2005. 'Phonological processing of speech and sign in the deaf brain'. Poster presented at the Cognitive Neuroscience Society, March, in New York.

Paulesu, E., et al. 2001. Dyslexia: Cultural diversity and biological unity. *Science* 291, no. 5511: 2165–7.

Porpodas, C.D. 1999. Patterns of phonological and memory processing in beginning readers and spellers of Greek. *Journal of Learning Disabilities* 32: 406–16.

Port, R. 2006. The graphical basis of phones and phonemes. In *Second language speech learning: The role of language experience in speech perception and production*, ed. M. Munro, and O. Schwen-Bohm. Amsterdam: John Benjamins.

Price, C.J., M.-L. Gorno-Tempini, K.S. Graham, N. Biggio, A. Mechelli, K. Patterson, and U. Noppeney. 2003. Normal and pathological reading: Converging data from lesion and imaging studies. *NeuroImage* 20, suppl. 1: S30–S41.

Price, C.J., and E. McCrory. 2005. Functional brain imaging studies of skilled reading and developmental dyslexia. In *The science of reading: A handbook*, ed. M.J. Snowling and C. Hulme, 473–96. Oxford: Blackwell.

Pugh, K. 2006. A neurocognitive overview of reading acquisition and dyslexia across languages. *Developmental Science* 9: 448–50.

Richardson, U., P.H.T. Leppänen, M. Leiwo, and H. Lyytinen. 2003. Speech perception of infants with high familial risk for dyslexia differ at the age of 6 months. *Developmental Neuropsychology* 23: 385–97.

Richardson, U., J. Thomson, S.K. Scott, and U. Goswami. 2004. Auditory processing skills and phonological representation in dyslexic children. *Dyslexia: An International Journal of Research and Practice* 10, no. 3: 215–33.

Rumsey, J.M., B. Horwitz, B.C. Donohue, K. Nace, J.M. Maisog, and P. Andreason. 1997. Phonological and orthographic components of word recognition: A PET rCBF study. *Brain* 120: 739–59.

Sandler, W., and D. Lillo-Martin. 2006. *Sign language and linguistic universals*. Cambridge: Cambridge University Press.

Sauseng, P., J. Bergmann, and H. Wimmer. 2004. When does the brain register deviances from standard word spellings? An ERP study. *Cognitive Brain Research* 20: 529–32.

Schneider, W., E. Roth, and E. Ennemoser. 2000. Training phonological skills and letter knowledge in children at-risk for dyslexia: A comparison of three kindergarten intervention programs. *Journal of Educational Psychology* 92: 84–95.

Share, D., and I. Levin. 1999. Learning to read and write in Hebrew. In *Learning to read and write: A cross-linguistic perspective*. In *Cambridge Studies in Cognitive and Perceptual Development*, ed. M. Harris and G. Hatano, 89–111. New York: Cambridge University Press.

Shaywitz, B.A., et al. 2002. Disruption of posterior brain systems for reading in children with developmental dyslexia. *Biological Psychiatry* 52, no. 2: 101–10.

———. 2007. Age-related changes in reading systems of dyslexic children. *Annals of Neurology* 61: 363–70.

Shaywitz, S.E., and B.A. Shaywitz. 2005. Dyslexia (specific reading disability). *Biological Psychiatry* 57: 1301–9.

Simos, P.G., J.I. Breier, J.M. Fletcher, E. Bergman, and A.C. Papanicolaou. 2000. Cerebral mechanisms involved in word reading in dyslexic children: A magnetic source imaging approach. *Cerebral Cortex* 10: 809–16.

Simos, P.G., et al. 2002. Dyslexia-specific brain activation profile becomes normal following successful remedial training. *Neurology* 58: 1203–13.

———. 2005. Early development of neurophysiological processes involved in normal reading and reading disability: A magnetic source imaging study. *Neuropsychology* 19, no. 6: 787–98.

Siok, W.T., C.A. Perfetti, Z. Jin, and L.H. Tan. 2004. Biological abnormality of impaired reading is constrained by culture. *Nature* 43: 71–6.

Snowling, M.J. 2000. *Dyslexia*. Oxford: Blackwell.

Stein, J., and V. Walsh. 1997. To see but not to read: The magnocellular theory of dyslexia. *Trends in Neuroscience* 20: 147–52.

Stuart, M. 2006. Teaching reading: Why start with systematic phonics teaching? *Psychology of Education Review* 30, no. 2: 6–17.

Stuart, M., and M. Coltheart. 1988. Does reading develop in a sequence of stages? *Cognition* 30: 139–81.

Szücs, D., and U. Goswami. 2007. Educational Neuroscience: Defining a new discipline for the study of mental representations. *Mind, Brain and Education* 1, no. 3: 114–27.

Tan, L.H., A.R. Laird, K. Li, and P.T. Fox. 2005. Neuroanatomical correlates of phonological processing of Chinese characters and alphabetic words: A meta-analysis. *Human Brain Mapping* 25, no. 1: 83–91.

Thomson, J.M., T. Baldeweg, and U. Goswami. 2005. 'Developmental trajectories ofauditory perception in dyslexia: an ERP study'. Poster presented at the 1st Course, International School on Mind, Brain and Education, July, in Sicily, Italy.

Turkeltaub, P.E., L. Gareau, D.L. Flowers, T.A. Zeffiro, and G.F. Eden. 2003. Development of neural mechanisms for reading. *Nature Neuroscience* 6, no. 6: 767–73.

Wimmer, H. 1996. The nonword reading deficit in developmental dyslexia: Evidence from children learning to read German. *Journal of Experimental Child Psychology* 61: 80–90.

Ziegler, J.C. 2005. Do differences in brain activation challenge universal theories of dyslexia? *Brain and Language* 98: 341–3.

Ziegler, J.C., and U. Goswami. 2005. Reading acquisition, developmental dyslexia, and skilled reading across languages: a psycholinguistic grain size theory. *Psychological Bulletin* 131, no. 1: 3–29.

———. 2006. Becoming literate in different languages: similar problems, different solutions. *Developmental Science* 9, no. 5: 429–53.

How should educational neuroscience conceptualise the relation between cognition and brain function? Mathematical reasoning as a network process

Sashank Varma and Daniel L. Schwartz

Stanford University, Stanford, CA, USA

Background: There is increasing interest in applying neuroscience findings to topics in education.

Purpose: This application requires a proper conceptualisation of the relation between cognition and brain function. This paper considers two such conceptualisations. The *area focus* understands each cognitive competency as the product of one (and only one) brain area. The *network focus* explains each cognitive competency as the product of collaborative processing among multiple brain areas.

Sources of evidence: We first review neuroscience studies of mathematical reasoning – specifically arithmetic problem-solving and magnitude comparison – that exemplify the area focus and network focus. We then review neuroscience findings that illustrate the potential of the network focus for informing three topics in mathematics education: the development of mathematical reasoning, the effects of practice and instruction, and the derailment of mathematical reasoning in dyscalculia.

Main argument: Although the area focus has historically dominated discussions in educational neuroscience, we argue that the network focus offers a complementary perspective on brain function that should not be ignored.

Conclusions: We conclude by describing the current limitations of network-focus theorising and emerging neuroscience methods that promise to make such theorising more tractable in the future.

Introduction

The relationship between education and neuroscience has been the subject of productive debate (Ansari and Coch 2006; Blakemore and Frith 2005; Bruer 1997; Byrnes and Fox 1998; Geake 2004; Goswami 2006; Varma, McCandliss and Schwartz in press). We supplement this discussion by describing two approaches to explaining how the brain gives rise to cognitive competence, and how they might contribute to educational thinking.

One appeal of cognitive neuroscience is that it is a 'place-based'. The topology of the brain yields the prospect of a spatial map that ties functions to areas. The place-based grounding of neuroscience theories makes them different from psychological theories, which are cast in terms of more abstract constructs like schemas, IQ and identity. It is literally possible to search databases by brain area to see which tasks cause them to activate – without ever entering a psychological keyword (e.g., Laird, Lancaster, and Fox 2005).

Figure 1 depicts two dominant approaches for understanding the place-based nature of cognition. The *area focus* typifies earlier theorising in cognitive neuroscience, and continues to characterise discussions in educational neuroscience. It decomposes cognition into a set of tasks and maps them to brain areas in a one-to-one fashion. Said differently, it seeks to identify *the* brain area that activates most selectively for each task competency. In contrast, the *network focus* explains task competency as the product of coordination among multiple brain areas. Network-focus research typically builds upon pioneering area-focus research that has identified initial landmarks. It expands the unit of analysis from the functioning of individual brain areas to the co-functioning of networks of brain areas.

Our concern is that the area focus currently dominates discussions in educational neuroscience, and it risks inappropriate inferences for improving educational practice. The one-to-one mapping of competencies to brain areas easily leads to the conclusion that students just need to exercise one part of their brain to develop or remediate a skill. It also naturally leads to the complaint that 'knowing where it sits in the brain does not tell us anything useful'. The problem with area-focus reasoning is that most tasks that educators care about are complex and multifaceted (especially compared with those studied by cognitive neuroscientists). These tasks are likely to map to brain areas in a many-to-many fashion. Said another way, most tasks activate multiple brain areas, and conversely most brain areas activate for multiple tasks. Moreover, the same task can be accomplished by different networks depending on experience (Tang et al. 2006). This paper argues that exclusively adopting an area focus risks the uptake of educational neuroscience in a seductive but premature form, and that a complementary network focus should also be emphasised. It grounds the argument primarily in the content area of mathematics.

This paper has the following structure. It first describes the area focus and illustrates its application to topics in mathematics education. Much of the discussion centres on two brain areas: intraparietal sulcus (IPS) and angular gyrus (AG). These areas are shown in Figure 2, along with a number of other areas that are mentioned below. Next, the area focus is incrementally broadened into the network focus through a broader consideration

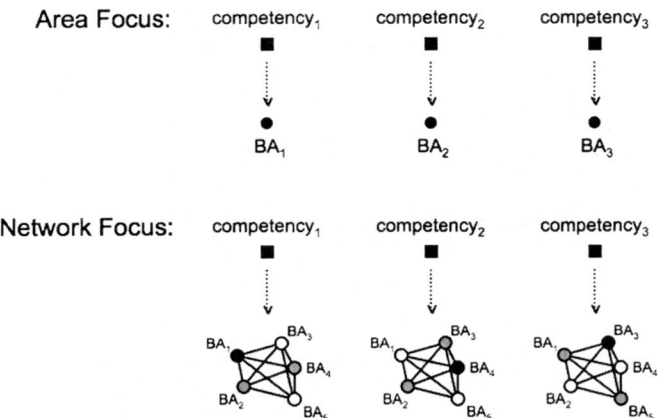

Figure 1. The area focus and network focus. The darker a circle, the more a brain area (**BA**) contributes to a competency.

of neuroscience findings on mathematical reasoning. Finally, the value of the network focus is illustrated by applying it to three topics in mathematics education: the development of mathematical reasoning, the effects of practice and instruction and the derailment of mathematical reasoning in dyscalculia.

The area focus for mathematical reasoning

The area focus has thus far dominated discussions in educational neuroscience. One reason for this dominance is that the methods of neuroscience have historically been well suited for isolating the brain areas necessary for a given ability. For example, in the nineteenth century, Broca encountered a patient with intact receptive language but impaired expressive language. Although the patient could comprehend language, he could only produce the utterance 'tan'. An autopsy revealed a lesion to a single brain area (left inferior frontal gyrus). Broca localised the expressive language competency to this area. A few years later, Wernicke applied the same logic to localise the receptive language competency to a different area (left posterior superior temporal gyrus). Another example of an area focus on brain function is the work conducted by the neurosurgeon Penfield in the early twentieth century. He electrically stimulated the brains of awake patients and observed their responses and impairments. A famous result of this research was the homunculi – topographical maps of somatosensory and motor cortex where adjacent brain areas coded sensation and action for adjacent regions of the body.

The area focus has been the dominant way to understand the results of neuroimaging experiments. Perhaps the most popular technique is functional magnetic resonance imaging (fMRI). When neurons fire, they make metabolic demands, consuming local stores of glucose and oxygen. This brings a haemodynamic response to replenish these

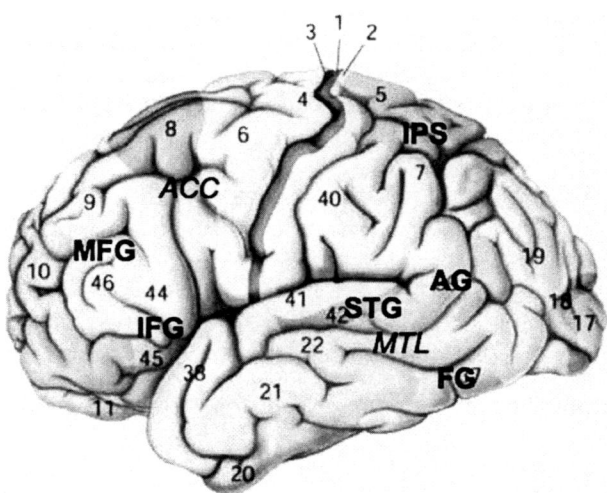

Figure 2.   Important brain areas for mathematical reasoning: intraparietal sulcus (IPS), angular gyrus (AG), Broca's area/inferior frontal gyrus (IFG), Wernicke's area/prosterior superior temporal gyrus (STG), fusiform gyrus (FG), medial temporal lobe/hippocampus (MTL), middle frontal gyrus (MFG) and anterior cingulate cortex (ACC). Lateral areas (i.e., near the outside of the brain) are labelled in bold, medial areas (i.e., near the centre of the brain) in italics. The numbers are according to Brodmann's scheme.

stores: the vascular system carries oxygenated blood to the region via arteries and carries deoxygenated blood away from the region via veins. Oxygenated blood and deoxygenated blood have different magnetic susceptibilities. As a result, differences in their relative concentration in a region produce differences in the magnetic resonance signal emanating from that region, and these differences can be used to generate images. As this brief description makes clear, fMRI is a rather indirect measure of neuronal activity: it registers the vascular response to metabolic activity in support of neuronal activity (for a more comprehensive description of fMRI see Huettel, Song, and McCarthy 2004). fMRI is popular because it can non-invasively measure activity in behaving brains, and because it provides good spatial resolution (i.e., each picture element has a volume in the order of 10 cubic millimeters) and acceptable temporal resolution (i.e., an image can be acquired every second or so).

The design and analysis of fMRI experiments have historically depended on the use of *tight subtractions*.[1] Participants complete two nearly identical tasks (e.g., naming digits versus naming letters). The fMRI scan produces a map of activation across the brain for each task. The map will include activation in areas of little theoretical interest, for example, due to moving the eyes or pressing a response button. To remove this 'noise', researchers subtract the activation map of the control task (e.g., letter naming) from the activation of the target task (e.g., digit naming). This leaves only the activation due to the competency of interest (e.g., accessing number). Over the past 15 years, thousands of fMRI experiments have used tight subtractions to map competencies to brain areas in a one-to-one manner.

In addition to the availability of suitable methods, another allure of the area focus is that it can be straightforwardly applied to understand the neural bases of complex forms of cognition. For example, consider the mathematical competency of being able to reason about numbers as magnitudes (Case et al. 1997) – what is also called 'number sense' (Dehaene 1997) and understanding 'numerosity' (Butterworth 2005; Landerl, Bevan, and Butterworth 2004). The area focus asks which brain area implements this competency. Neuroscientists have pursued this question by capitalising on the symbolic distance effect (SDE) – the finding that the time taken to compare two digits decreases as the distance between them increases; for example, people are faster to judge which of 1 versus 9 is larger than to judge which of 1 versus 3 is larger (Moyer and Landauer 1967). The SDE is commonly interpreted as evidence that people reason about numerical magnitudes using a 'mental number line' that is psychophysically scaled, so that, much like perceptual discriminations (e.g., loudness and softness), values that are closer together on the number line are harder to discriminate than values that are far apart. Neuroscientists have used the SDE to identify the 'numerical magnitude area' of the brain. A representative study is by Pinel et al. (2004). Participants compared pairs of digits, judging which was greater. A handful of brain areas showed an increase in activation that paralleled the increasing response times for closer comparisons. Most prominent among them was IPS.[2] From an area focus, this is evidence that this brain area is the primary correlate of the numerical magnitude competency – that it is the seat of the mental number line.

The area focus can also provide insights about individual differences, which present a natural bridge from neuroscience to education (Kosslyn and Koenig 1992). The area focus describes a deficit as a dysfunction of the brain area that implements the relevant ability. This is a variant of the reasoning that Broca and Wernicke applied to understand language impairments, augmented with the assumption that a structurally intact area can be rehabilitated by exercising it through repeated practice of the relevant task. The application of this reasoning produced the biggest success story in educational neuroscience to date, the remediation of one subtype of dyslexia. In a representative study, Eden et al. (2004) used fMRI to first identify the networks of brain areas recruited by typical readers and those with dyslexia. Dyslexic readers showed reduced activation in AG, which has been implicated in mapping orthography to phonology. Next, the dyslexic readers participated in a program developed by educational researchers for remediating phonological difficulties. Post-test fMRI scans revealed that successful remediation was associated with increased activation in AG. From an area-focus approach, this 'weak' brain area had been 'strengthened'.

An area focus is currently being applied to understand dyscalculia, the mathematical analog of dyslexia. Dyscalculia is defined as scoring in the lowest 5% (or so) on tests of mathematical achievement relative to age, education level and intelligence (Butterworth 2005). This is a coarse clinical definition, and dyscalculia is likely a blanket term that includes multiple subtypes. Molko et al. (2003) applied the logic of the area approach to understand the mathematical impairment of a relatively homogeneous group of dyscalculics – those with Turner syndrome. They focused on the mathematical competency of arithmetic problem-solving – the ability to compute or retrieve the answers to addition and subtraction problems (and, in other experiments, multiplication and division problems) where the operands are small positive integers. They capitalised on the *problem size effect*: the finding that the time to solve problems with large operands (e.g., 8 + 9) is slower than the time to solve problems with small operands (e.g., 4 + 3) (Ashcraft 1992). Stanescu-Cosson et al. (2000) had previously identified a neural analog of the problem size effect in normal adults, finding that operand size correlates positively with activation in IPS. Molko et al. (2003) found that patients with dyscalculia failed to show a problem size effect in IPS (or any other brain area).[3] An area-focus interpretation of this finding is that under-activation of IPS in this group of dyscalculics is correlated with their impaired arithmetic problem-solving. The next logical step would be a training study to exercise this 'mental muscle', with the expected result that performance would improve and IPS activation would come to resemble that of people without dyscalculia.

The network focus for mathematical reasoning

Although an area focus is important for initially mapping the functional terrain of the brain, it ultimately presents an oversimplified view of the neural bases of mathematical reasoning. That one area is necessary for a particular ability does not imply that it is sufficient. A broader consideration of neuroimaging studies reveals that many mathematical competencies are better viewed as emergent products of networks of brain areas. As a corollary, some impairments of mathematical reasoning may be better viewed as breakdowns in network function; consequently, remediation may require exercises that *coordinate* areas rather than strengthen them in isolation.

The network focus has been a minor theme in neuroscience theorising for decades. An early example comes from Lashley, who incrementally removed portions of rats' brains to identify 'the memory area'. His conclusion was that no such area existed, and that the rat brain instead worked by *mass action*: the more that was removed, the more performance declined. Though it ultimately proved to be an untenable account of memory, the proposed distribution of function served as a useful counterweight to the area focus. Another early example of a focus on network function is Luria (1966), who observed that focal brain lesions often impair not a single competency, but rather a range of competencies, some more than others. More recently, Mesulam (1990) has argued that attention and language are better understood as the products of partially overlapping, large-scale cortical networks. In this view, most competencies are implemented by multiple areas, and most areas contribute to multiple competencies.

fMRI studies are increasingly focusing on the network of brain areas that activates for a given task, rather than the single area that activates most selectively. For example, consider the neural bases of face recognition. Early studies found evidence that fusiform gyrus (an area in inferior temporal cortex) selectively activates for processing faces when activation associated with the processing of other visual categories, such as houses, is subtracted away (Kanwisher, McDermott, and Chun 1997). This led to the label 'fusiform face area' and the concomitant assumption that the ability to discriminate faces had enough survival value that the human brain evolved a dedicated area. However, subsequent studies revealed that fusiform gyrus activates not just for faces, but also for other visual categories such as houses and furniture, though to a lesser extant (Ishai et al. 1999). Conversely, other inferior temporal areas that activate selectively for other visual categories also activate for faces, though to a lesser extant. In this way, an initial area-based understanding of face recognition has been articulated into a more nuanced network-based understanding. The remainder of this section describes a similar (and ongoing) shift, where an initial area-based understanding of arithmetic problem-solving is being refined into a network-based understanding.

Early neuroimaging studies of adults found selective activation of IPS when subtracting single-digit operands. Within the area focus, this was interpreted as evidence that IPS implements the subtraction competency. Because other researchers had found activation in this area during visuospatial processing, Dehaene et al. (2003) proposed that subtraction problems are solved by imagining and moving along a mental number line. In contrast, early studies of multiplication found selective activation of AG. This was interpreted as evidence that this area implements the multiplication competency. Because other researchers had found AG activation during retrieval of phonological information, Dehaene et al. (2003) proposed that multiplication is performed by look-up in a verbally coded, mental multiplication table. In this way, the area focus made sense of early neuroimaging studies – subtraction involves visuospatial processing and multiplication verbal processing.

Though simple and elegant, the area focus can miss potential complexities revealed by a network focus. For example, Lee (2000) had participants solve subtraction and multiplication problems in the scanner and found network effects. Multiple brain areas activated more for subtraction than multiplication; IPS was one, but it was not the only one. Conversely, multiple brain areas activated more for multiplication than subtraction; AG was one, but it was not the only one. These results suggested that mathematical competencies might be better understood as the products of networks of brain areas, not single brain areas.

In the preceding examples, researchers used tight subtractions: activation during multiplication was subtracted from activation during subtraction, and vice versa. By definition, each activation peak was associated with one, and only one, arithmetic operation. This led naturally to the inference of *independent* brain areas in the case of Dehaene et al. (2003) and *independent* (i.e., non-overlapping) networks of brain areas in the case of Lee (2000). Other studies have used 'loose subtractions' to isolate activation patterns. In a loose subtraction, activation from a relatively low-level control condition, such as viewing a fixation cross, is subtracted from activations during the experimental conditions of interest. The result is a more complete picture of the network recruited by each experimental condition. Studies employing loose subtractions reveal that subtraction and multiplication activate a *common* network of brain areas, although they activate different areas to different degrees. For example, Chochon et al. (1999) subtracted activation when viewing a fixation cross from activation during subtraction and multiplication respectively. They found that subtraction activated a network of brain areas, one that included IPS. Critically, they found that multiplication activated almost the same network. This network included IPS, although it was activated less intensely.

Duffau et al. (2002) conducted a neurosurgical study of a patient with a tumour in AG. Before removing the tumour, electro-stimulation was used to map competencies within AG. Among other tasks, the patient solved different kinds of arithmetic problems. Electrical stimulation was directly applied to different sites within AG, so it was possible to see which competencies were disrupted. Consistent with an area focus, the researchers found a multiplication site within AG. Critically, they also found a subtraction site in the same brain area, as well as a site common to both operations. These results suggest that it is a mistake to narrowly construe AG as *the* multiplication area. Rather, it is a component of a larger arithmetic network, and it plays a role not just in multiplication, but also in subtraction (and likely other aspects of mathematical reasoning as well).

These network findings indicate that the mapping of behaviour to the brain is more complex than that suggested by an area focus and frequently communicated to educators and educational researchers. The different pictures of arithmetic painted by the area and network approaches are important for education because they may have different implications for how best to teach. The area focus suggests that subtraction should be taught using spatial referents such as number lines to capitalise on the functional specialisation of IPS; and that multiplication should be taught verbally, for example, by rehearsing times tables, to recruit AG. In contrast, the network approach is consistent with instruction that targets the development of number sense (Baroody 1985). Children should be given opportunities to integrate different meanings and operations of number by engaging in activities that yield coordinated networks (Case et al. 1997). Note that this prescription does not preclude development of a mental number line, nor large doses of mathematical fact memorisation. However, it does suggest that a number line representation is not sufficient for achieving flexible subtraction competence, and memorisation is not sufficient for achieving flexible multiplication competence. As we describe below, there is a place for both meaning and memorisation in arithmetic.

Using the network approach to understand topics in mathematics education

The area focus currently dominates how neuroscience findings are packaged for educational researchers. As a result, the potential of the network focus remains largely untapped. This section illustrates this potential. It applies the network focus to three topics of interest to mathematics education: the development of mathematical reasoning, the effects of practice and instruction and the derailment of mathematical reasoning in dyscalculia. The examples show how a network focus can refine the broad-stroke neuroscience models one might use to explain educationally relevant phenomena.

Qualitative shifts underlying continuous behavioural changes

Developmental neuroscientists were among the first to adopt a network focus (e.g., Johnson et al. 2002). Consider the development of the understanding that digits name quantities or magnitudes. The SDE (i.e., the difference in response times for comparing near digits versus far digits) is indicative of whether people have developed an interpretation of number that includes its magnitude interpretation. In a cross-sectional study, Sekuler and Mierkiewicz (1977) documented that the SDE (i.e., the difference in response times for comparing near digits versus far digits) is present as early as kindergarten and decreases continuously into adulthood (but never completely). The area focus predicts that this continuous change in the degree of the SDE should be accompanied by a continuous change in the activation of IPS.[4] Ansari et al. (2005) tested this prediction by having adults and 10-year-old children make numerical comparisons. The adults showed an SDE in a network of brain areas that included IPS, replicating prior studies. Critically, for the children, an activation pattern differentiating near versus far comparisons was not observed in IPS, though it was observed in other brain areas. In the case of numerical magnitude, a continuous developmental change at the behavioural level belies a qualitative shift at the neural level.

Another example, from the domain of arithmetic problem solving, comes from a cross-sectional study by Rivera et al. (2005). Children between the ages of 8 and 19 solved simple addition and subtraction problems. Although accuracy was constant across development, there was a continuous improvement in solution speed with age. Recall that the area focus predicts that a continuous change in behavioural performance with development should be accompanied by a continuous change in the activation level of the corresponding neural correlate. However, the Rivera et al. (2005) results were more consistent with the network focus. Some areas of the arithmetic network were more active early in development. These areas have been implicated in domain-general forms of cognition (i.e., prefrontal areas associated with controlled processing and executive function and medial temporal areas associated with declarative long-term memory). Other areas became more active with development, including those known to be associated with visuospatial processing (IPS) and verbal processing (AG). These are more domain-specific forms of cognition. Once again, a continuous developmental change at the behavioural level – faster addition and subtraction – is better understood as a qualitative shift in the underlying network, in this case, reflecting a transition from domain-general to domain-specific processing.[5] This qualitative shift raises the question of whether educational activities should change over time to help students move from early domain-general processing to later domain-specific processing. Whether a constant dose of thought-provoking problems is the best way to encourage the shift, or whether practising the same types of problems repeatedly better encourages the shift, are interesting empirical questions raised by a network focus.

Effects of memorisation and strategy training

The network approach helps clarify the effects of practice on mathematical reasoning. Delazer et al. (2003) trained participants on complex multiplication problems, where a two-digit operand is multiplied by a one-digit operand. They were then scanned as they solved the same problems they had studied, plus a set of new problems of similar difficulty. This design makes it possible to identify the learning effects of memorising specific mathematical facts through practice versus computing them. Activation in AG (and some other areas) increased for the trained problems, suggesting that answers were being accessed from a verbal store. In addition, activation in IPS (and some other areas) decreased for trained problems, suggesting that less computation was performed for familiar problems. One interpretation of these results is that practice produced a shift in the arithmetic network that reflected a transition from a more computational visuospatial strategy to a more retrieval-based verbal strategy for the trained problems.

The Delazer et al. (2003) study is important because it addresses the effects of practice, an issue of interest to mathematics education. Delazer et al. (2005) took the next step in a study that examined the effects of pure memorisation versus learning an algorithm for computing solutions. They taught participants a novel arithmetic operation using two kinds of instruction. The memorisation group memorised the answers to problems with specific operands. They never learned how to compute the operation. By contrast, the strategy group was taught an algorithm for computing the answer given the same operands. Both groups then solved familiar and novel problems in the scanner.

The results showed that participants in the memorisation condition organised one network of brain areas to perform the operation and participants in the strategy condition another. For example, the memorisation network included AG, which has been implicated in the retrieval of verbally coded knowledge, whereas the strategy network included the anterior cingulate cortex, which has been implicated in controlled cognitive processing. This difference is important for two reasons. First, it is a difference at the brain level that matters at the behavioural level, and is thus relevant for education. The network organised by participants in the strategy condition supported transfer to novel problems (78% accuracy), whereas the network organised by participants in the memorisation condition did not (15% accuracy). Memorisation and calculation strengthen different networks rather than strengthening the same one, and thus the network analysis helps explain the differential effects of memorising versus learning to calculate. A second important contribution of this study for the prospects of educational neuroscience is that it demonstrates that fMRI can be used to study the consequences of instruction delivered outside the scanner over a relatively long period of time.

Dyscalculia as network under-activation

Recall that Molko et al. (2003) contrasted a group of normal controls with a group of dyscalculics as they solved addition problems. The critical finding was that normal controls displayed a problem size effect in the activation of IPS, whereas dyscalculics did not. Although the results of this study are comprehensible from an area focus, those of a more recent study of dyscalculia are better understood from a network focus. Kucian et al. (2006) imaged a group of dyscalculics and a group of normal controls as they performed a range of mathematical tasks. In one task, approximate addition, they found under-activation of the entire arithmetic network in the dyscalculic group relative to the normal control group. The implicated areas included bilateral IPS, inferior frontal gyrus, middle frontal gyrus and anterior cingulate cortex. These results suggest that understanding dyscalculia will require focusing on both the dysfunction of individual brain areas and the dysfunction of networks of brain areas.[6] It is an open question of what kinds of instruction may be able to organise a dysfunctioning network (as opposed to a dysfunctioning brain area, which we saw above in the dyslexia example: Eden et al. 2004)? We return to this question below.

Conclusion

This paper has considered two approaches to understanding the relationship between cognition and brain function. The area focus maps cognitive competencies to brain areas in a one-to-one fashion. The network focus understands each cognitive competency as the emergent product of information processing in a network of brain areas. Although the area focus has historically dominated discussions, we argued the network focus offers a complementary perspective on brain function that educational neuroscience should not ignore.

Two of the examples presented above bring the area focus and network focus into particularly sharp contrast. The first concerns the arithmetic problem-solving of typical adults. Initial studies adopted an area focus. Their findings suggested that subtraction selectively activates IPS, and thus involves visuospatial processing, whereas multiplication selectively activates AG, and thus involves verbal processing (Dehaene et al. 2003). Subsequent studies adopted a network focus. In contrast, they found evidence for a common arithmetic network whose component brain areas are taxed differently by different operations (Chochon et al. 1999; Duffau et al. 2002; Lee 2000). The second example where both the area focus and network focus have been adopted is dyscalculia. Although the study of this impairment is still in its infancy, an early study by Molko et al. (2003) adopted an area approach. It found that a neural correlate of dyscalculia was dysfunction of IPS. By contrast, the more recent study by Kucian et al. (2006) adopted a network focus. It found under-activation not of a single brain area, but rather the entire arithmetic network. The network-focus conclusions are consistent with the views of many in mathematics education (Baroody 1985; Case et al. 1997), namely that arithmetic problem-solving is the product of an interrelated set of mathematical competencies, and that the failure to properly coordinate these competencies results in poor mathematical achievement. For this reason, we expect the network focus to become increasingly important as educational neuroscience matures.

We conclude by describing the current limitations of network focus theorising and emerging neuroscience methods that promise to make such theorising more tractable in the future. An important limitation of the network focus for education is that it posits a complex, many-to-many mapping of mathematical competencies to brain areas. This makes it difficult to make predictions about the effects of network function and dysfunction, and therefore to draw implications for questions of interest to educational researchers. By contrast, the area focus maps mathematical competencies to brain areas in a one-to-one fashion, with a deficit in a particular competency understood as a dysfunction of the corresponding brain area. This has a natural educational implication: to design instruction that 'strengthens' that 'weak' area, presumably improving performance. Although this approach has had a few limited successes (e.g., Eden et al. 2004), its prospects are ultimately limited by the fact that the brain is *not* carved at the same functional joints that make sense at the behavioural level. Rather, brain areas appear to be specialised for lower-level functions, and it is only through their organisation in large-scale networks that these functions coalesce into mathematical competencies that matter at the behavioural level, and are thus of interest to educational researchers.

However, there are methods that make network-style theorising more tractable. They should enable studies that ask how brain areas become connected and coordinated in networks, as when children learn to coordinate cardinal and ordinal conceptions of quantity (Case et al. 1997). One example is *functional connectivity analysis*, which looks for correlated activity in different brain areas during task performance (e.g., Friston 1994). The inference is that correlated brain areas are communicating as part of a large-scale network. For example, Büchel, Coull, and Friston (1999) found that learning gains were associated not with changes in the activation of a single brain area, but rather with increases in correlated activity among brain areas. Functional connectivity analysis may be useful for understanding the network-wide under-activations in dyscalculia documented by Kucian et al. (2006). This deficit may be better understood as a dysfunction of how well brain areas communicate with, and therefore co-activate, one another. Another promising neuroscience method is *diffusion tensor imaging* (DTI), which directly images the anatomical connections – the white matter tracts – over which brain areas communicate (e.g., Le Bihan et al. 2001). The potential of DTI to inform topics in education is illustrated by a recent study by Niogi and McCandliss (2006), who found that the integrity of left temporo-parietal white-matter tracts is correlated with reading ability in elementary school children. Future functional connectivity and DTI studies of mathematical reasoning, literacy and other forms of cognition of interest to educational neuroscientists promise to benefit from a network focus.

Acknowledgements

We thank the editor and two anonymous reviewers for helpful comments. This material is based upon work supported by the National Science Foundation under grants REC 0337715 and SLC-0354453. Opinions, findings and conclusions or recommendations expressed in this material are those of the authors and do not necessarily reflect the views of the National Science Foundation.

Notes

1. The use of subtraction has declined over the years as other experimental designs and methods of analysis have been developed. We describe two of these advancements in the 'Conclusion' section.
2. Pinel et al. (2004) also had participants compare stimuli along physical dimensions, such as size and luminance. These comparisons also produced SDEs in IPS. Comparisons of numerical magnitude and physical size activated roughly the same peak coordinates in IPS, whereas the comparisons of physical luminance activated different peak coordinates, though in the same area.
3. The dyscalculic patients did show a behavioural problem size effect, but it was exaggerated relative to normal controls, suggesting use of a different strategy (e.g., verbal counting versus magnitude-based processing).
4. Whether the change is an increase or decrease in activation depends on one's conception of what develops (Poldrack 2000). If one believes that representations get richer, then the prediction is increasing activation. If one believes that representations are shaped or tuned (i.e., made more efficient), then the prediction is decreasing activation.
5. There are other ways to interpret this shift. Rivera et al. (2005) favour an attentional interpretation, from more controlled to more automatic processing. Importantly, this interpretation is also a network explanation.
6. The Kucian et al. (2006) results do not strictly compel a network interpretation. It is possible to interpret them from an area focus if one assumes that the dyscalculia is not a homogeneous deficit, but rather is composed of multiple subtypes; and that each subtype is associated with dysfunction of a single competency, and thus a single brain area.

References

Ansari, D., and D. Coch. 2006. Bridges over troubled waters: Education and cognitive neuroscience. *Trends in Cognitive Sciences* 10: 146–51.

Ansari, D., N. Garcia, E. Lucas, K. Hamon, and B. Dhital. 2005. Neural correlates of symbolic number processing in children and adults. *NeuroReport* 16: 1769–73.

Ashcraft, M.H. 1992. Cognitive arithmetic: A review of data and theory. *Cognition* 44: 75–106.

Baroody, A.J. 1985. Mastery of basic number combinations: Internalization of relationships of facts? *Journal for Research in Mathematics Education* 16: 83–98.

Blakemore, S.-J., and U. Frith. 2005. *The learning brain: Lessons for education.* Oxford: Blackwell.

Bruer, J.T. 1997. Education and the brain: A bridge too far. *Educational Researcher* 26: 4–16.

Büchel, C., J.T. Coull, and K.J. Friston. 1999. The predictive value of changes in effective connectivity for human learning. *Science* 283: 1538–41.

Butterworth, B. 2005. Developmental dyscalculia. In *Handbook of mathematical psychology*, ed. J.I.D. Campbell, 455–67. Hove: Psychology Press.

Byrnes, J.P., and N.A. Fox. 1998. The educational relevance of research in cognitive neuroscience. *Educational Psychology Review* 10: 297–342.

Case, R., Y. Okamoto, S. Griffin, A. McKeough, C. Bleiker, B. Henderson, K.M. Stephenson, R.S. Siegler, and D.P. Keating. 1997. The role of central conceptual structures in the development of children's thought. *Monographs of the Society for Research in Child Development* 61: 1–295.

Chochon, F., L. Cohen, P.F. Van de Moortele, and S. Dehaene. 1999. Differential contributions of the left and right inferior parietal lobules to number processing. *Journal of Cognitive Neuroscience* 11: 617–30.

Dehaene, S. 1997. *The number sense.* New York: Oxford University Press.

Dehaene, S., M. Piazza, P. Pinel, and L. Cohen. 2003. Three parietal circuits for number processing. *Cognitive Neuropsychology* 20: 487–506.

Delazer, M., F. Domahs, L. Bartha, C. Brenneis, A. Lochy, T. Trieb, and T. Benke. 2003. Learning complex arithmetic – an fMRI study. *Cognitive Brain Research* 18: 76–88.

Delazer, M., A. Ischebeck, F. Domahs, L. Zamarian, F. Koppelstaetter, C.M. Siedentopf, L. Kaufmann, T. Benke, and S. Felber. 2005. Learning by strategies and learning by drill – evidence from an fMRI study. *NeuroImage* 25: 838–49.

Duffau, H., D. Denvil, M. Lopes, F. Gasparini, L. Cohen, L. Capelle, and R. Van Effenterre. 2002. Intraoperative mapping of the cortical areas involved in multiplication and subtraction: An electrostimulation study in a patient with a left parietal glioma. *Journal of Neurology, Neurosurgery, and Psychiatry* 73: 733–8.

Eden, G.F., K.M. Jones, K. Cappell, L. Gareau, F.B. Wood, T.A. Zeffiro, N.A.F. Dietz, J.A. Agnew, and D.L. Flowers. 2004. Neural changes following remediation in adult developmental dyslexia. *Neuron* 44: 411–22.

Friston, K.J. 1994. Functional and effective connectivity in neuroimaging: A synthesis. *Human Brain Mapping* 2: 56–78.

Geake, J. 2004. Cognitive neuroscience and education: Two-way traffic or one-way street? *Westminster Studies in Education* 27: 87–98.

Goswami, U. 2006. Neuroscience and education: From research to practice. *Nature Reviews Neuroscience* 7: 406–13.

Huettel, S.A., A.W. Song, and G. McCarthy. 2004. *Functional magnetic resonance imaging.* Sunderland, MA: Sinauer.

Ishai, A., L.G. Ungerleider, A. Martin, J.L. Schouten, and J.V. Haxby. 1999. Distributed representation of objects in the human ventral visual pathway. *Proceedings of the National Academy of Science USA* 96: 9379–84.

Johnson, M.H., H. Halit, S.J. Grice, and A. Karmiloff-Smith. 2002. Neuroimaging of typical and atypical development: A perspective from multiple levels of analysis. *Development and Psychopathology* 14: 521–36.

Kanwisher, N., J. McDermott, and M.M. Chun. 1997. The fusiform face area: A module in human extrastriate cortex specialized for face perception. *Journal of Neuroscience* 17: 4302–11.

Kosslyn, S.M., and O. Koenig. 1992. *Wet mind: The new cognitive neuroscience.* New York: The Free Press.

Kucian, K., T. Loenneker, T. Dietrich, M. Dosch, E. Martin, and M. Von Aster. 2006. Impaired neural networks for approximate calculation in dyscalculic children: A functional MRI study. *Behavioral and Brain Functions* 2: 1–17.

Laird, A.R., J.L. Lancaster, and P.T. Fox. 2005. BrainMap: The social evolution of a functional neuroimaging database. *Neuroinformatics* 3: 65–78.

Landerl, K., A. Bevan, and B. Butterworth. 2004. Developmental dyscalculia and basic numerical capacities: A study of 8–9-year-old students. *Cognition* 93: 99–125.

Le Bihan, D., J.-F. Mangin, C. Poupon, C.A. Clark, S. Pappata, N. Molko, and H. Chabriat. 2001. Diffusion tenor imaging: Concepts and applications. *Journal of Magnetic Resonance Imaging* 13: 534–46.

Lee, K.-M. 2000. Cortical areas differentially involved in multiplication and subtraction: A functional magnetic resonance imaging study and correlation with a case of selective acalculia. *Annals of Neurology* 48: 657–61.

Luria, A.R. 1966. *Higher cortical functions in man.* London: Tavistock.

Mesulam, M.-M. 1990. Large-scale neurocognitive networks and distributed processing for attention, language and memory. *Annals of Neurology* 28: 597–613.

Molko, N., A. Cachia, D. Rivière, J.-F. Mangin, M. Bruandet, D. Le Bihan, L. Cohen, and S. Dehaene. 2003. Functional and structural alterations of the intraparietal sulcus in a developmental dyscalculia of genetic origin. *Neuron* 40: 847–58.

Moyer, R.S., and T.K. Landauer. 1967. Time required for judgments of numerical inequality. *Nature* 215: 1519–20.

Niogi, S.N., and B.D. McCandliss. 2006. Left lateralized white matter microstructure accounts for individual differences in reading ability and disability. *Neuropsychologia* 44: 2178–88.

Pinel, P., M. Piazza, D. Le Bihan, and S. Dehaene. 2004. Distributed and overlapping cerebral representations of number, size, and luminance during comparative judgments. *Neuron* 41: 983–93.

Poldrack, R.A. 2000. Imaging brain plasticity: Conceptual and methodological issues – a theoretical review. *NeuroImage* 12: 1–13.

Rivera, S.M., A.L. Reiss, M.A. Eckert, and V. Menon. 2005. Developmental changes in mental arithmetic: Evidence for increased specialization in the left inferior parietal cortex. *Cerebral Cortex* 15: 1779–90.

Sekuler, R., and D. Mierkiewicz. 1977. Children's judgments of numerical inequality. *Child Development* 48: 630–3.

Stanescu-Cosson, R., P. Pinel, P.-F. Van de Moortele, D. Le Bihan, L. Cohen, and S. Dehaene. 2000. Understanding dissociations in dyscalculia: A brain imaging study of the impact of number size on the cerebral networks for exact and approximate calculation. *Brain* 123: 2240–55.

Tang, Y., W. Zhang, K. Chen, S. Feng, Y. Ji, J. Shen, E.M. Reiman, and Y. Liu. 2006. Arithmetic processing in the brain shaped by cultures. *Proceedings of the National Academy of Science USA* 103: 10775–80.

Varma, S., B.D. McCandliss, and D.L. Schwartz. In press. Scientific and pragmatic challenges for bridging education and neuroscience. *Educational Researcher*.

Dyscalculia: neuroscience and education

Liane Kaufmann

Clinical Department of Pediatrics IV, Section Neuropediatrics, Innsbruck Medical University, Austria

Background: Developmental dyscalculia is a heterogeneous disorder with largely dissociable performance profiles. Though our current understanding of the neurofunctional foundations of (adult) numerical cognition has increased considerably during the past two decades, there are still many unanswered questions regarding the developmental pathways of numerical cognition. Most studies on developmental dyscalculia are based upon adult calculation models which may not provide an adequate theoretical framework for understanding and investigating developing calculation systems. Furthermore, the applicability of neuroscience research to pedagogy has, so far, been limited.

Purpose: After providing an overview of current conceptualisations of numerical cognition and developmental dyscalculia, the present paper (1) reviews recent research findings that are suggestive of a neurofunctional link between fingers (finger gnosis, finger-based counting and calculation) and number processing, and (2) takes the latter findings as an example to discuss how neuroscience findings may impact on educational understanding and classroom interventions.

Sources of evidence: Finger-based number representations and finger-based calculation have deep roots in human ontology and phylogeny. Recently, accumulating empirical evidence supporting the hypothesis of a neurofunctional link between fingers and numbers has emerged from both behavioural and brain imaging studies.

Main argument: Preliminary but converging research supports the notion that finger gnosis and finger use seem to be related to calculation proficiency in elementary school children. Finger-based counting and calculation may facilitate the establishment of mental number representations (possibly by fostering the mapping from concrete non-symbolic to abstract symbolic number magnitudes), which in turn seem to be the foundations for successful arithmetic achievement.

Conclusions: Based on the findings illustrated here, it is plausible to assume that finger use might be an important and complementary aid (to more traditional pedagogical methods) to establish mental number representations and/or to facilitate learning to count and calculate. Clearly, future prospective studies are needed to investigate whether the explicit use of fingers in early mathematics teaching might prove to be beneficial for typically developing children and/or might support the mapping from concrete to abstract number representations in children with and without developmental dyscalculia.

Introduction

Arithmetic learning disorders (developmental dyscalculia) denote circumscribed and outstanding difficulties in the acquaintance of arithmetic skills. Importantly, dyscalculia is not a unitary concept and the associated cognitive profiles might vary widely between and within individuals (for overviews see Kaufmann and Nuerk 2005; Wilson and Dehaene 2007). With an estimated prevalence of 3% to 7%, developmental dyscalculia is about as frequent as developmental dyslexia (APA 1994) and, similar to dyslexia, persists into adulthood if untreated (Shalev and Gross-Tsur 2001). It is widely agreed that dyscalculia is a highly familial disorder (the risk for siblings of children suffering from dyscalculia is five to ten times higher than in the general population: Shalev et al. 2001). Though developmental dyscalculia may present as a single-deficit disorder (e.g., core deficit of 'number sense': Landerl, Bevan, and Butterworth 2004; for a review see Wilson and Dehaene 2007), many affected children exhibit associated cognitive problems, both within and outside the numerical domain (for respective overviews see Kaufmann and Nuerk 2005; Shalev and Gross-Tsur 2001). The frequent occurrence of comorbidities coincides with recent findings reporting a substantial genetic overlap between various developmental learning disorders such as dyscalculia, dyslexia and attention-deficit hyperactivity disorder (Plomin, Kovas, and Haworth 2007). Nevertheless, dyscalculia research is much younger than dyslexia research and most of what is known is derived from adult studies involving mature brain systems. Hence, our current understanding of the behavioural manifestations and (neuro)cognitive foundations of developmental dyscalculia remains incomplete.

The main aim of this paper is to demonstrate the need to go beyond adult calculation models when attempting to account for the peculiarities of developing brain systems and developmental disorders such as developmental dyscalculia. Moreover, the paper aims to illustrate how findings from brain imaging studies may inform educational understanding and even classroom instructions/interventions. The example discussed here concerns the association between fingers and numbers which has not been considered explicitly in adult calculation models, but which nevertheless – as we shall see – plays a fundamental role in learning to count and calculate. Butterworth (1999) merits reward for initiating the greatly renewed neuroscientific interest in the potential importance of finger use for the acquaintance of numerical skills. According to Butterworth (1999), fingers are convenient and natural tokens to represent number magnitudes which are intuitively used by young children when learning the verbal counting sequence and/or when executing first simple additions and subtractions (see also Butterworth 2005). Moreover, even adults may use finger-based back-up strategies. Thus, it may be argued that finger-based number representations and finger-based calculation are deeply rooted in human ontology and phylogeny. Indeed, converging evidence from brain imaging findings corroborate the latter notion of a neurofunctional link between fingers and numbers (Andres, Seron, and Olivier 2007; Kaufmann et al. 2008; Sato et al. 2007; Thompson et al. 2004). Before presenting the respective findings and their potential educational implications in more detail, we present a brief overview of current conceptualisations of typical and atypical developmental pathways of numerical cognition (as derived from neuropsychology and neuroscience).

Current conceptualisations of numerical cognition and developmental dyscalculia

Infants as young as 4–6 months are able to make number-based discriminations and even exhibit additive expectation behaviour (up to set sizes of three objects: Wynn 1992, 1995).

Larger set size discriminations may be mastered, too, provided the ratio of the to-be-compared object sets is large enough (e.g., babies may discriminate 8 from 16, but not 8 from 12 objects: Xu and Spelke 2000; and even 16 from 32 objects, but not 16 from 24: Xu, Spelke, and Goddard 2005). Interestingly, when continuous stimulus characteristics such as contour length and total filled area are controlled, 6-month-old babies can successfully discriminate large, but not small, numerosities (Xu, Spelke, and Goddard 2005; see also Xu 2003). Consequently, Xu (2003) and Xu, Spelke, and Goddard (2005) propose the existence of two core systems for number magnitude representation in infants: one mediating small (exact) and the other supporting large (approximate) numerical magnitudes (see also Feigenson, Dehaene, and Spelke 2004; for a similar distinction in mature brain systems see Dehaene and Cohen 1997).

In verbal individuals (i.e., children who have begun to master language) numerical concepts emerge as soon as children use counting to refer to objects. Upon entering formal education, most typically developing children demonstrate a rudimentary understanding of number relations, are able to count up to 20 and may even master simple additions and subtractions verbally when allowed to use their fingers or other reference objects. Thus, upon starting school (at age 6 in most European countries), most children already have acquired some verbal counting and calculation skills, which are considered to be the foundations for establishing more advanced school mathematics.

But what do we know about the development of these number processing and calculation skills during infancy and preschool years? How do children develop quantity knowledge and number representations? Which neurocognitive processes and mechanisms come into play when children gradually acquire abstract (symbolic) number representations during formal schooling? And which difficulties – within and outside the numerical domain – accompany developmental dyscalculia? These latter questions are at the core of current research attempts aiming to delineate the developmental pathways of numerical cognition. However, dyscalculia research is relatively young and, in the absence of an empirically validated *developmental* calculation model, many neuropsychological studies targeted at gaining a better understanding of the developmental pathways of typical and/ or atypical numerical cognition rest on adult calculation models (e.g., Dehaene and Cohen 1995; Dehaene et al. 2003; McCloskey, Caramazza, and Basili 1985).

According to popular adult calculation models, number processing and calculation is multi-componential and the components constituting the calculation system are thought to be modularly organised (Dehaene and Cohen 1995; McCloskey, Caramazza, and Basili 1985). The modularity assumption derives from adult cognitive neuropsychology which considers double dissociations as evidence for a modular architecture of (neuro)cognitive systems and/or mental representations (Shallice 1988). For example, a double dissociation is present if cognitive ability A is preserved while ability B is deficient in one individual, and in another individual the opposite pattern emerges (i.e., preserved ability B and deficient ability A). Probably the most popular adult calculation model is the so-called 'triple code model' of numerical cognition (Dehaene and Cohen 1995) postulating three modularly organised but interrelated calculation components (i.e., analogue magnitude representation, auditory verbal word frame, visual Arabic number form), each of which is thought to be supported by distinct brain regions. A consistent finding in the adult literature concerns the key role of the intraparietal sulcus (IPS) for number magnitude processing (for an overview see Dehaene et al. 2003; Hubbard et al. 2005).

Up to now, numerous (mostly adult) studies have been published taking the Dehaene calculation model (Dehaene and Cohen 1995, 1997) as a starting-point upon which to test their hypotheses (for respective reviews see Dehaene et al. 2003; Hubbard et al. 2005; see also for developmental issues Kaufmann and Nuerk 2005; Wilson and Dehaene 2007). However, because of crucial differences between developing and mature brain systems, adult models may not provide adequate theoretical frameworks for investigating developmental disorders (see Bishop 1997; Karmiloff-Smith 1992). For example, double dissociations in developmental disorders need not necessarily reflect the presence of modularly organised neurofunctional networks (Karmiloff-Smith 1997; Pennington 2006). Indeed, double dissociations have been observed in non-modular cognitive architectures as well (i.e., in the case of dyslexia the double dissociation between phonological and surface dyslexia; Harm and Seidenberg 1999; Plaut 1995).

Behavioural studies suggest that number magnitude discrimination abilities may be an innate capacity inherent to infants (Wynn 1992, 1995; Xu and Spelke 2000; Xu 2003) and even non-human species (Brannon and Roitman 2003), yet the question arises whether number magnitude processing is supported by identical brain regions in infants and adults. Although present developmental findings are consistent with a neurofunctional link between intraparietal regions and number magnitude processing, there is some controversy regarding the age-dependency of intraparietal (IPS) involvement in the formation of arithmetical skills. While some findings reveal similar activations in intraparietal regions extending across different ages during number magnitude processing (symbolic number processing: Temple and Posner 1998; non-symbolic number processing: Cantlon et al. 2006; Temple and Posner 1998), other studies suggest that the functional specialisation of the IPS for number magnitude processing increases with age (non-symbolic number processing: Ansari and Dhital 2006; symbolic number processing: Ansari et al. 2005; Kaufmann et al. 2005, 2006; Rivera et al. 2005). These conflicting results may partly be explained by methodological differences between studies, making a direct comparison across studies (and paradigms) difficult. Alternatively, and as mentioned already above, one may claim that adult models (resting on modularity assumptions) are not apt to account for the complexity of developing brain systems.

Developmental dyscalculia: single- or multiple-deficit views?

Upon adopting Dehaene's modularly organised adult calculation model (Dehaene and Cohen 1995; Dehaene et al. 2003), some researchers propose that the neurocognitive underpinnings of developmental dyscalculia are best conceptualised as a (single) core deficit of 'number sense' (e.g., Butterworth 2005; Landerl, Bevan, and Butterworth 2004). The core deficit hypothesis implies that children diagnosed with developmental dyscalculia display specific difficulties to mentally represent and manipulate (non-symbolic) number magnitudes. Consequently, the core deficit of 'number sense' is thought to be related to a malfunctioning of intraparietal brain regions (i.e., the horizontal segment of the intraparietal sulcus (HIPS), which is supposed to mediate number magnitude processing according to Dehaene et al. 2003). In an excellent review, Wilson and Dehaene (2007) revisit this strong single-deficit view of developmental dyscalculia by arguing that the core deficit of 'number sense' may be only one of several possibly underlying deficits. According to Wilson and Dehaene (2007), other potential subtypes of dyscalculia – each of which being supported by distinct brain regions – may rest on (1) deficient verbal symbolic representations (manifesting themselves as arithmetic fact retrieval difficulties); (2) deficient executive functions (hampering fact retrieval as well as complex calculation); or (3) deficient spatial attention (leading to impaired quick recognition of small numerosities and possibly negatively affecting non-symbolic and symbolic number manipulations). Thus, Wilson and Dehaene (2007) propose that the behavioural characteristics of developmental dyscalculia – and their neurocognitive underpinnings – might vary substantially between individuals, thus seriously questioning a strong single-deficit view.

Another argument challenging the single-deficit view of developmental dyscalculia is the observation that many children exhibiting problems in learning to count and calculate also have difficulties in other cognitive domains. Even within the arithmetical domain, children display quite distinguishable performance profiles (both at an intraindividual and interindividual level of analysis: Dowker 2005). Consequently, various attempts to classify developmental dyscalculia at a behavioural level have been undertaken (e.g., Geary 2000; Temple 1989, 1991; Von Aster 2000). As early as 1991, Temple reported a double dissociation between arithmetic fact retrieval (e.g., number fact knowledge such as 3 × 5) and procedural knowledge ('knowing how' to solve a complex arithmetic problem) in developmental dyscalculia (however, see for a critical discussion of double dissociations in developmental disorders Pennington 2006). The distinction between arithmetic fact and procedural knowledge was first acknowledged in the adult calculation model proposed by McCloskey and colleagues (1985). Likewise, the theoretical foundations for Geary's (2000) and Von Aster's (2000) efforts to classify developmental dyscalculia were grossly based on the Dehaene calculation model (Dehaene and Cohen 1995). A commonality of the latter classification attempts is their effort to further differentiate developmental dyscalculia according to specific performance profiles, which in turn imply the existence of distinct single cognitive deficits. And yet, according to Pennington (2006), any attempt to link these deficits to one – and only one – underlying neuroanatomical (and/or genetic) underpinning is likely to fail. Rather, developmental dyscalculia should be regarded as a complex and dynamic developmental disorder (for similar views of dyslexia and attention-deficit hyperactivity disorder see Bishop 1997; Pennington 2006). Interestingly, and consistent with the latter view, recent findings of quantitative genetic research report a substantial genetic overlap between quite diverse cognitive (dis)abilities such as reading, language and arithmetic (Plomin and Kovas 2005; for a review see also Plomin et al. 2007). This genetic overlap may partly explain the repeatedly reported high incidence of comorbidity of developmental disorders such as dyslexia, dyscalculia and attentional disorders (for a review see Kaufmann and Nuerk 2005), and furthermore, may also partly account for the considerable diversity of neurocognitive performance within one developmental disorder (in our case, dyscalculia).

To summarise, although single-deficit models (e.g., core deficit of 'number sense': Butterworth 2005; Landerl, Bevan, and Butterworth et al. 2004; 'number fact dyscalculia': Temple 1991) are presently predominant in the neuroscientific literature – probably because they are simpler and hence more testable – multiple-deficit models of developmental dyscalculia seem to better fit our current understanding of the complex nature of developmental disorders.[1] Thus, a change of paradigms from a modular and single-deficit view towards a dynamic, process-oriented and multiple-deficit view seems to be essential for the development of empirically validated *developmental* calculation models, as well as for the production of mathematics curricula meeting children's neurocognitive development (i.e., maturation-dependent readiness to grasp number-based concepts and skills) and the generation of tailored dyscalculia intervention programmes.

Neuroscience and education: the case of developmental dyscalculia

A frequent criticism of brain imaging studies involving learning is their restricted applicability to education and classroom interventions. Indeed, the great majority of neuroscientific research – including the realm of numerical cognition and/or developmental dyscalculia – is targeted at basic research. As neuroscience is a rather young discipline, which is further tightly connected to recent technological advances (and which underwent and still continues to undergo dramatic changes within very short periods), early respective studies are hardly comparable to more recent ones (regarding both methodological and practical issues). Further, it is important to acknowledge that significant activations reported in imaging studies reveal brain regions modulating a specific task which is *not* equivalent to regions being *necessary* to process the task at hand. In addition, and partly because of methodological and technical constraints, experimental paradigms generally focus on islets of skills rather than *learning processes and mechanisms*, the latter being a greater focus of interest for educational researchers and classroom teachers. In general, imaging studies are only as good as the behavioural paradigms they are implementing. Hence, the development of adequate behavioural paradigms should be based on a sophisticated understanding of the interplay between neurocognitive (including genetic) and pedagogical factors determining typical and atypical trajectories within particular cognitive domains (i.e., in our case, numerical cognition).

Recently, researchers of both disciplines (i.e., neuroscience and education) are slowly becoming aware of the urgent need to ameliorate communication and to foster common research efforts. The latter focus is reflected in continuously appearing scientific articles devoted to the topic of 'neuroscience and education' (Ansari and Coch 2006; Fawcett and Nicolson 2007; Goswami 2004; Szucs 2005), as well as in the newly founded scientific journal *Mind, Brain and Education*, whose first issue was compiled in 2007. Furthermore, Fawcett and Nicolson (2007) request the establishment of a new discipline of 'pedagogical neuroscience'. These authors emphasise that diagnostic efforts based on behavioural and/or cognitive symptoms are not sufficient to contribute to our understanding of complex developmental disorders. Taking developmental dyslexia as an example, Fawcett and Nicolson (2007) stress the need to develop brain-based theories (by employing genetic and brain-based diagnostic methods), which eventually may advance not only our understanding of developmental disorders, but also lead to tailored interventions. Hence, there is a clear need for research designs being specifically targeted at the educational implications of neuroscience research. In order to accomplish the latter goal, educational experts must share their expertise in pedagogy, and neuroscience researchers must develop ecological paradigms that are capable of investigating cognitive processes and learning mechanisms instead of circumscribed skills.

Below, I present a brain imaging study conducted at Innsbruck Medical University which was aimed at making a first step towards bridging the gap between neuroscience and education (Kaufmann et al. 2008). Although the main aim of our study was to elucidate the link between non-symbolic numerical and spatial processing, here I will focus on the numerical task only and the potential implications for educational sciences that arose from studying it. The numerical task required participants to make simple number comparisons. Stimuli were pictures of two hands, each hand showing a different finger pattern (e.g., the right hand raising three fingers, the left one two fingers). Thus, our experimental paradigm provoked finger-based counting/number discriminations and these helped us address some questions regarding the association between fingers and numbers. Before discussing the results, I briefly present the relevant literature that led us to formulate our working hypotheses.

Fingers and numbers

Empirical evidence for a link between fingers and numbers is derived from developmental behavioural studies (Fayol, Barrouillet, and Marinthe 1998; Gracia-Baffaluy and Noel 2008; Landerl, Bevan, and Butterworth 2004; Noel 2005; Sato and Lalain 2008), patient studies (Gerstmann syndrome: Gerstmann 1940; developmental Gerstmann syndrome: Benson and Geschwind 1970; Suresh and Sebastian 2000) and brain imaging studies (fMRI: Simon et al. 2002; Thompson et al. 2004; TMS: Andres, Seron, and Olivier 2007; Roux et al. 2003; Rusconi, Walsh, and Butterworth 2005; Sato et al. 2007).

Probably the earliest report of a neurofunctional association between fingers (i.e., finger discrimination) and number processing was provided in 1940 by Gerstmann, who described a patient with a right posterior parietal lesion accompanied by symptoms combining finger agnosia (difficulties to recognise and discriminate fingers), acalculia (outstanding calculation problems), right–left disorientation and agraphia (impaired writing: see Benton 1997; for descriptions of developmental Gerstmann syndromes see, e.g., Benson and Geschwind 1970; Suresh and Sebastian 2000).[2]

Associations between fingers (finger gnosis) and calculation skills have also been reported in developmental behavioural studies. For instance, in typically developing preschool children, neuropsychological test scores (including finger recognition and finger discrimination) were found to be a good predictor of calculation skills one year later (Fayol, Barrouillet, and Marinthe 1998). Furthermore, the findings of Noel (2005) revealed that finger gnosis seems to be a specific predictor for numerical abilities and further suggest that the link between finger gnosis and arithmetic is not restricted to tasks relying on finger-based magnitude representations (but rather encompasses a wide range of number processing tasks). Noel (2005) argues that the latter findings are best explained by the anatomical vicinity of brain regions supporting finger gnosis and those mediating number magnitude processing and calculation.

Consistent with the latter suggestion, results of a functional magnetic resonance imaging (fMRI) study (Simon et al. 2002) revealed neighbouring and partly overlapping activations in posterior parietal brain regions for quite diverse abilities such as arithmetic and goal-directed hand movement (grasping/pointing), among others. In particular, brain regions supporting grasping (postcentral gyrus and anterior IPS) were found to border those mediating calculation (in and around the (H)IPS: see Simon et al. 2002, figure 2; for a review, see also Hubbard et al. 2005).

Further corroborating the notion of a neurofunctional link between finger use and number processing are the results of a repetitive transcranial magnetic resonance (rTMS) study revealing that both finger movements and number magnitude judgements (Arabic digits) are disrupted by left parietal stimulation in adults (i.e., angular gyrus: Rusconi, Walsh, and Butterworth 2005; see also Roux et al. 2003). Finally, two recent TMS studies assessing corticospinal excitability in hand muscles are suggestive of (1) a special role of right-hand muscles (left hemisphere) for small numerals (1–4, which were interpreted as reflecting culturally acquainted embodied finger counting strategies: Sato et al. 2007; for consistent behavioural findings see Sato and Lalain 2008); and (2) of a link between hands (but not arms and/or legs) and enumeration (numbers and letters: Andres, Seron, and Olivier 2007). Interestingly, upon investigating 16- and 17-year-old adolescents with and without a diagnosis of developmental dyscalculia (DD), Soltész and collaborators (2007) report that electrophysiological responses upon performing a simple, one-digit number comparison task were not comparable between the two groups (though both groups displayed comparable behavioural performance on this task). In particular, and most interestingly, relative to their non-DD peers, individuals with DD displayed very specific neuropsychological performance profiles being characterised by preserved mental rotation and body part knowledge (among others) but deficient performance on mental finger rotation and finger knowledge. Thus, the latter results provide the first evidence that, in DD (or some groups of DD), deficient finger knowledge may be associated with atypical brain mechanisms for performing a basic numerical task.

The latter results are clearly exciting, but it has to be noted that all respective brain imaging studies were performed on adults and hence provide information about mature brain systems only. Indeed, despite converging behavioural evidence for an association between finger gnosis and numerical skills in children (e.g., Fayol, Barrouillet, and Marinthe 1998; Noel 2005; Sato et al. 2007), respective developmental brain imaging studies are so far lacking. This gap in the research provided our motivation for conducting a developmental fMRI study that required 8-year-old children (and young adults) to make number magnitude judgements. More specifically, we presented stimuli that consisted of pictures of two hands representing different numerosities (i.e., finger patterns). Participants were asked to indicate, by pressing a button, which hand displayed more fingers. Thus, by provoking finger-based comparison strategies, the experimental paradigm required participants to make (non-symbolic) numerical classifications.

Besides number discriminations, participants were asked to make spatial and colour discriminations, too.[3] As a thorough discussion of this research clearly goes beyond the scope of the present paper, results and educational implications presented here will focus on the following questions: (1) do elementary school children and adults recruit identical brain regions upon solving a simple number comparison task (provoking finger-based number representations)?; and (2) is there a neurofunctional link between finger-based number representations and counting/number comparison, and if so, is there an age-related change in cerebral activation patterns related to finger-based magnitude extraction?

Results revealed highly interesting findings. Behaviourally, children and adults performed at ceiling upon making number classifications (99.3% correct). However, compared with adults, children were significantly slower (746 ms and 1017 ms respectively), although response latency patterns for different types of pairs were again comparable between age groups. In particular, both age groups were significantly quicker to classify distant relative to adjacent number pairs (i.e., displaying shorter response latencies upon comparing 1 versus 5 relative to 1 versus 2: children $p < 0.05$; adults $p < 0.001$). The latter reaction time phenomenon has been coined the 'distance effect' and is thought to reflect the integrity of the mental number line (Dehaene 1991). Thus, the behavioural data suggest that the task was performed flawlessly by both groups (as reflected by very high accuracy rates) and, moreover, children and adults alike processed number magnitude all the way down to the semantic numerical level (as reflected by the presence of the distance effect). However, a different picture emerged regarding brain activation patterns. In particular, activation patterns in response to non-symbolic number processing were clearly distinguishable between children and adults. Relative to adults, children recruited additional brain areas in lateral portions of anterior IPS, as well as in adjacent regions of the right post- and precentral gyrus upon making finger-based number magnitude classifications. Most interestingly, the latter regions were found to be deactivated in adults. We interpret our findings as being suggestive of an age-dependent neurofunctional link between areas supporting finger use and non-symbolic number processing (Kaufmann et al. 2008). Importantly, the latter findings imply that even in the case of comparable behavioural performance between children and adults, brain activation patterns need not be identical across age.

Potential educational implications of our findings

Our findings provide evidence for an age-dependent link between finger-use and number processing that is not only interesting for neuroscience and numerical cognition research, but may also have significant implications for educational research and even classroom teaching. The demonstration of age-dependent activation differences when solving a simple number comparison task (which was performed at ceiling by both children and adults) emphasises the importance of going beyond behavioural (performance) issues. Our findings highlight the potential benefit of incorporating our steadily increasing understanding of developing neurofunctional systems into efforts to design adequate and timely pedagogical curricula. In other words, although children may succeed in solving a particular (numerical) task as well as adults in terms of performance, children may need to put more effort than adults into orchestrating the brain regions associated most closely with the task. For instance, as illustrated above, a simple number comparison task requires 8-year-old children – but less so adults – to recruit brain areas supporting finger use, thus revealing age-dependent processing mechanisms at a neural level.

With respect to classroom teaching, the implications of the latter findings are straightforward. As brain areas mediating finger use might be co-activated whenever children need to access mental number representations, it does not seem advisable to forbid children using their fingers upon performing arithmetic problems. Rather, educators and teachers could take advantage of the fact that fingers may serve as concrete embodied tokens to represent number magnitude. Moreover, fingers mirror the base-10 number system, and moreover, are readily available to be used as back-up strategies. Thus, it is plausible to expect that the consistent use of fingers could positively affect the formation of mental number representations (by facilitating the mapping from concrete non-symbolic quantity knowledge to abstract symbolic number processing) and thus also the acquisition of calculation skills. Indeed, preliminary evidence supporting the latter claim comes from a recent intervention study demonstrating that training finger gnosis significantly improves arithmetic performance in 1st-graders (Gracia-Baffaluy and Noel 2008). Moreover, a prospective study of elementary school children revealed a predominance of split-five calculation errors (i.e., solutions deviating ± 5 from the correct result: Domahs, Krinzinger, and Willmes 2008). The latter authors interpreted their findings as reflecting 'failure to keep track of "full hands" in counting or calculation' (abstract: Domahs, Krinzinger, and Willmes 2008). Interestingly, with increasing age/schooling (i.e., grades 1 to 3) split-five errors decreased, thus suggesting that children's reliance on mental finger patterns (whole hand/five fingers) decreases with increasing schooling/calculation proficiency.

Finally, it is not far-fetched to argue that the explicit incorporation of finger-use in numeracy intervention programs could be beneficial for the establishment of mental number representations in children suffering from developmental dyscalculia. However, in the absence of respective empirical studies, the latter claim thus far remains speculative.

Last but not least, it is important to stress that finger knowledge is not the whole story to becoming good at maths. Rather, several other skills like abstract thinking (e.g., facilitating the mapping process between concrete and symbolic arithmetic), spatial skills (enabling the formation of a spatially oriented mental number line and, moreover, our understanding of the base-10 system of the Arabic number system), working memory (enabling us to manipulate quantities when solving arithmetic tasks, to monitor multi-step procedures, etc.), language proficiency (underlying counting routines and arithmetic fact retrieval, among others) are also considered to be important for becoming a proficient calculator. Each of the latter domains is likely to mediate the acquisition and/or application of arithmetic skills and future – preferably longitudinal – research endeavours are needed to elucidate their impact on the development of numerical cognition (for a review see Kaufmann and Nuerk 2005; Wilson and Dehaene 2007).

To summarise, although the study described here nicely demonstrates the usefulness of tailoring research questions to pedagogical demands with respect to a core numerical skill, there is a clear need for future (prospective) research designs targeted at more complex learning processes and mechanisms within the realm of numerical cognition. The most sensible way to approach – and possibly achieve – the latter goal is to intensively foster scientific communication and expertise between neuroscience and education.

Acknowledgement

The author was supported by the Austrian Science Foundation (grant number T286-B05).

Notes

1. A major disadvantage of 'complex models' is that they may render the verification and/or falsification of working hypotheses difficult (because they typically encompass many dependent and/or unknown variables). Hence, researchers aiming to assess complex developmental disorders are especially required to (1) formulate very clear-cut hypotheses from the outset; (2) carefully define selection criteria for their study populations; and (3) employ paradigms that have been found previously to be adequate (and testable) for the research questions of interest. Reasons for advocating 'complex models' are at least twofold: first, complex models readily acknowledge modulating cognitive abilities (within and outside the numerical domain: Kaufmann and Nuerk 2005; Wilson and Dehaene 2007); and second, complex models may lead to a better understanding of the link between mind (cognitive), brain (neurofunctional) and pedagogy (behavioural and educational factors) mediating the acquaintance of number processing and calculation skills.
2. Though Benton (1997) seriously questioned the entity of the syndrome by stressing that a substantial proportion of patients exhibit some, but not all four symptoms constituting the full Gerstmann syndrome, the Gerstmann syndrome has received increased interest recently.
3. Stimuli across the three tasks were identical (only instructions varying), thus enabling us to control for domain-general perceptual and response-bound processing mechanisms. The strict control of domain-general processing mechanisms is crucial in brain imaging research as the to-be-interpreted activation patterns should be attributable to task-relevant processing solely (or as far as possible). The latter endeavour is achieved by a subtraction method whereby the cerebral activation patterns obtained in response to a control task (which is preferably identical to the experimental task in all but the variable of interest, in our case, number processing) are subtracted from the activations obtained in response to the experimental task. According to the subtraction logic, only the task-relevant – hence domain-specific – activations should remain. In order to achieve the best possible match between experimental and control tasks, we created stimuli that could be used across all three task conditions. In particular, stimuli consisted of two simultaneously displayed children's hands with coloured thumbs. In half of the trials the palms of the two hands showed in the same direction, while in the other half they did not. The spatial task required participants to judge whether the palms of the two hands were showing in the same direction or not. Likewise, in the colour condition, individuals were asked to state whether the colours of the two thumbs were identical or not. The colour task served as a true control task. The spatial task was incorporated in the study because our main aim was to disentangle spatial and non-symbolic numerical processing (Walsh 2003; for a comprehensive review on the neurofunctional overlap between spatial and numerical processing see Hubbard et al. 2005).

References

American Psychiatric Association (APA). 1994. *Diagnostic and statistical manual for mental disorders: DSM IV*. Washington, DC: American Psychiatric Association.

Andres, M., X. Seron, and E. Olivier. 2007. Contribution of hand motor circuits to counting. *Journal of Cognitive Neuroscience* 19, no. 4: 563–76.

Ansari, D., and D. Coch. 2006. Bridges over troubled waters: Education and cognitive neuroscience. *Trends in Cognitive Sciences* 10, no. 4: 146–51.

Ansari, D., and B. Dhital. 2006. Age-related changes in the activation of the intraparietal sulcus during non-symbolic magnitude processing: An event-related functional magnetic resonance imaging study. *Journal of Cognitive Neuroscience* 1, no. 11: 1820–8.

Ansari, D., N. Garcia, E. Lucas, K. Hamon, and B. Dhital. 2005. Neural correlates of symbolic number processing in children and adults. *NeuroReport* 1, no. 16: 1769–73.

Benson, F., and N. Geschwind. 1970. Developmental Gerstmann syndrome. *Neurology* 20: 293–8.

Benton, A.L. 1997. Reflections on the Gerstmann syndrome. *Brain and Language* 4: 45–62.

Bishop, D.V. 1997. Cognitive neuropsychology and developmental disorders: Uncomfortable bedfellows. *Quarterly Journal of Experimental Psychology* 5, no. 4: 899–923.

Brannon, E.M., and J.D. Roitman. 2003. Nonverbal representation of time and number in animals and human infants. In *Functional and neural mechanisms of interval timing*, ed. W.H. Meck, 143–82. New York: CRC Press.

Butterworth, B. 1999. *The mathematical brain*. London: Macmillan.

———. 2005. The development of arithmetical abilities. *Journal of Child Psychology and Psychiatry* 46, no. 1: 3–18.

Cantlon, J.F., E.M. Brannon, E.J. Carter, and K.A. Pelphrey. 2006. Functional imaging of numerical processing in adults and 4-year-old children. *PLoS Biology* 4, no. 5: e125.

Dehaene, S., and L. Cohen. 1995. Towards an anatomical and functional model of number processing. *Mathematical Cognition* 1: 83–120.

———. 1997. Cerebral pathways for calculation: Double dissociation between rote verbal and quantitative knowledge of arithmetic. *Cortex* 33, no. 2: 219–50.

Dehaene, S., G. Dehaene-Lambertz, and L. Cohen. 1998. Abstract representation of numbers in the animal and human brain. *Nature Neuroscience* 21: 355–61.

Dehaene, S., M. Piazza, P. Pinel, and L. Cohen. 2003. Three parietal circuits for number processing. *Cognitive Neuropsychology* 20: 487–506.

Domahs, F., H. Krinzinger, and K. Willmes. 2008. Mind the gap between both hands: Evidence for internal finger-based number representations in children's mental calculation. *Cortex* 44: 359–67.

Dowker, A. 2005. *Individual differences in arithmetic: Implications for psychology, neuroscience and education*. Hove: Psychology Press.

Fawcett, A.J., and R.I. Nicolson. 2007. Dyslexia, learning, and pedagogical neuroscience. *Developmental Medicine and Child Neurology* 49: 306–11.

Fayol, M., P. Barrouillet, and C. Marinthe. 1998. Predicting arithmetical achievement from neuropsychological performance: A longitudinal study. *Cognition* 68: B63–B70.

Feigenson, L., S. Dehaene, and E. Spelke. 2004. Core systems of number. *Trends in Cognitive Sciences* 8, no. 7: 307–14.

Geary, D.C. 2000. From infancy to adulthood: The development of numerical abilities. *European Child and Adolescent Psychiatry* 9, no. 2: 11–16.

Gerstmann, J. 1940. Syndrome of finger agnosia, disorientation of right and left, agraphia and dyscalculia. *Archives of Neurology and Psychiatry* 44: 298–408.

Goswami, U. 2004. Neuroscience and education. *British Journal of Educational Psychology* 74: 1–14.

Gracia-Baffaluy, M., and M.-P. Noel. 2008. Does finger training increase numerical performance? *Cortex* 44: 368–75.

Harm, M.W., and M.S. Seidenberg. 1999. Phonology, reading acquisition, and dyslexia: Insights from connectionist models. *Psychological Review* 10, no. 3: 491–528.

Hubbard, E.M., M. Piazza, P. Pinel, and S. Dehaene. 2005. Interactions between number and space in parietal cortex. *Nature Reviews Neuroscience* 6: 435–48.

Karmiloff-Smith, A. 1992. *Beyond modularity: A developmental perspective in cognitive science*. Cambridge, MA: MIT Press/Bradford Books.

———. 1997. Crucial differences between developmental cognitive neuroscience and adult neuropsychology. *Developmental Neuropsychology* 13, no. 4: 513–24.

Kaufmann, L., F. Koppelstaetter, M. Delazer, C. Siedentopf, P. Rhomberg, S. Golaszewski, S. Felber, and A. Ischebeck. 2005. Neural correlates of distance and congruity effects in a numerical Stroop task: An event-related fMRI study. *NeuroImage* 25: 888–98.

Kaufmann, L., F. Koppelstaetter, C. Siedentopf, I. Haala, E. Haberlandt, L.-B. Zimmerhackl, S. Felber, and A. Ischebeck. 2006. Neural correlates of a number-size interference task in children. *NeuroReport* 1, no. 6: 587–91.

Kaufmann, L., and H.-C. Nuerk. 2005. Numerical development: Current issues and future perspectives. *Special issue: Brain and Number, Psychology Science* 47, no. 1: 142–70.

Kaufmann, L., S. Vogel, G. Wood, C. Kremser, M. Schocke, L.-B. Zimmerhackl, and J.W. Koten. 2008. A developmental fMRI study of non-symbolic numerical and spatial processing. *Cortex* 44: 376–85.

Landerl, K., A. Bevan, and B. Butterworth. 2004. Developmental dyscalculia and basic numerical capacities: A study of 8–9-year-old students. *Cognition* 93, no. 2: 99–125.

McCloskey, M., A. Caramazza, and A. D. Basili. 1985. Cognitive mechanisms in number processing and calculation: Evidence from dyscalculia. *Brain and Cognition* 4: 171–96.

Noel, M.-P. 2005. Finger gnosia: A predictor of numerical abilities in children? *Child Neuropsychology* 11: 413–30.

Pennington, B.F. 2006. From single- to multiple-deficit models of developmental disorders. *Cognition* 101, no. 2: 385–413.

Plaut, D. 1995. Double dissociations without modularity: Evidence from connectionist neuropsychology. *Journal of Clinical and Experimental Neuropsycholgy* 17: 291–321.

Plomin, R., and Y. Kovas. 2005. Generalist genes and learning disabilities. *Psychological Bulletin* 131: 592–617.

Plomin, R., Y. Kovas, and C.M.A. Haworth. 2007. Generalist genes: Genetic links between brain, mind, and education. *Mind, Brain, and Education* 1, no. 1: 11–19.

Rivera, S.M., A.L. Reiss, M.A. Eckert, and V. Menon. 2005. Developmental changes in mental arithmetic: Evidence for increased functional specialization in the left inferior parietal cortex. *Cerebral Cortex* 24, no. 1: 50–60.

Roux, F.E., S. Boetto, O. Sacko, F. Chollet, and M. Tremoulet. 2003. Writing, calculating, and finger recognition in the region of the angular gyrus: A cortical stimulation study of Gerstmann syndrome. *Journal of Neurosurgery* 99, no. 4: 716–27.

Rusconi, E., V. Walsh, and B. Butterworth. 2005. Dexterity with numbers: rTMS over left angular gyrus disrupts finger gnosis and number processing. *Neuropsychologia* 43, no. 11: 1609–24.

Sato, M., L. Cattaneo, G. Rizzolatti, and V. Gallese. 2007. Numbers within our hands: Modulation of corticospinal excitability of hand muscles during numerical judgment. *Journal of Cognitive Neuroscience* 19, no. 4: 684–93.

Sato, M., and M. Lalain. 2008. On the relationship between handedness and hand-digit mapping in finger counting. *Cortex* 44: 393–99.

Shalev, R.S., and V. Gross-Tsur. 2001. Developmental dyscalculia. *Pediatric Neurology* 24, no. 5: 337–42.

Shalev, R.S., O. Manor, B. Kerem, M. Ayali, N. Badichi, Y. Friedlander, and V. Gross-Tsur. 2001. Developmental dyscalculia is a familial learning disability. *Journal of Learning Disabilities* 34: 59–65.

Shallice, T. 1988. *From neuropsychology to mental structure.* New York: Cambridge University Press.

Simon, O., J.-F. Mangin, L. Cohen, D. Le Bihan, and S. Dehaene. 2002. Topographical layout of hand, eye, calculation, and language-related areas in the human parietal lobe. *Neuron* 33: 475–87.

Soltész, F., D. Szücs, J. Dékány, A. Márkus, and V. Csépe. 2007. A combined event-related potential and neuropsychological investigation of developmental dyscalculia. *Neuroscience Letters* 417: 181–6.

Suresh, P.A., and S. Sebastian. 2000. Developmental Gerstmann syndrome: A distinct clinical entity of learning disabilities. *Pediatric Neurology* 22: 267–78.

Szucs, D. 2005. Teachers can substantially inform cognitive psychological and cognitive neuroscience research. *Journal of the Professional Association of Teachers of Students with Specific Learning Disabilities* November: 4–7.

Temple, C.M. 1989. Digit dyslexia: A category-specific disorder in developmental dyscalculia. *Cognitive Neuropsychology* 6: 93–116.

———. 1991. Procedural dyscalculia and number fact dyscalculia: Double dissociation in developmental dyscalculia. *Cognitive Neuropsychology* 8, no. 2: 155–76.

Temple, E., and M. Posner. 1998. Brain mechanisms of quantity are similar in 5-year-olds and adults. *Proceedings of the National Academy of Science USA* 95: 7836–41.

Thompson, J.C., D.F. Abbott, K.J. Wheaton, A. Syngeniotis, and A. Puce. 2004. Digit representation is more than just hand waving. *Cognitive Brain Research* 21: 412–17.

Von Aster, M. 2000. Developmental cognitive neuropsychology of number processing and calculation: Varieties of developmental dyscalculia. *European Child and Adolescent Psychiatry* 9, no. suppl. 2: 41–57.

Walsh, V. 2003. Cognitive neuroscience: Numerate neurons. *Current Biology* 13: 447–8.

Wilson, A., and S. Dehaene. 2007. Number sense and developmental dyscalculia. In *Human behavior, learning, and the developing brain: Atypical development*, ed. D. Coch, G. Dawson, and K. Fischer, 212–38. New York: Guilford Press.

Wynn, K. 1992. Addition and subtraction by human infants. *Nature* 358: 749–50.

———. 1995. Origins of numerical knowledge. *Mathematical Cognition* 1: 35–60.

Xu, F. 2003. Numerosity discrimination in infants: Evidence for two systems of representation. *Cognition* 89, no. 1: B15–B25.

Xu, F., and E.S. Spelke. 2000. Large number discrimination in 6-month-old infants. *Cognition* 74: B1–B11.

Xu, F., E.S. Spelke, and S. Goddard. 2005. Number sense in human infants. *Developmental Science* 8, no. 1: 88–101.

What are the implications of neuroscience for musical education?

Lauren Stewart[a] and Aaron Williamon[b]

[a]Department of Psychology, Goldsmiths, University of London, UK; [b]Centre for Performance Science, Royal College of Music, London, UK

Background: In this paper, we consider music education in a broad sense – not merely pertaining to the development of exceptional levels of artistry in talented performers, but also to notions of musical listening and appreciation enjoyed by the casual listener.
Purpose: This review cannot be exhaustive, but aims to illustrate what we already know about the neuroscience of how music is perceived, appreciated, learned and performed, and the implications that this knowledge has for music education in this broadly defined sense.
Design and methods: Extant studies from across the fields of neuroscience, psychology, education and music were surveyed using mainstream Internet databases (e.g., PubMed), as well as specific Internet cites promoting interdisciplinary exchange among musicians and scientists (e.g., Music and Science Online: http://www.science.rcm.ac.uk). The result is a review of some 50 studies from across this relatively young field.
Conclusions: To date, examples of tangible, practical advice from neuroscience that can be applied directly to musical learning and performance are relatively scant. However, the field is growing rapidly, and collaborations between musicians and scientists are becoming more common. We argue that the scope for neuroscience research to inform and shape musical education is ripe for development, particularly when musicians and scientists work together to address questions of musical relevance with scientific rigour.

How musical am I?

The idea of what counts as musical ability is heavily influenced by culture. In societies where communal music-making forms an important aspect of everyday life, such as in the Venda people of South Africa, making music is as natural as breathing, and the notion that someone could be 'unmusical' would be considered absurd (Blacking 1995). In Western cultures, by contrast, music is typically performed by individuals or small groups for the appreciation of large audiences, giving rise to the view that musical ability is the preserve of the formally trained. A study conducted by Cuddy and colleagues (2005) showed that 14% of undergraduates in an American university believed themselves to be tone-deaf, even though testing of their musical perception using a standardised battery (Peretz, Champod, and Hyde 2003) showed that only a small percentage of these individuals scored outside the normal range.

For many, the ability to sing in tune represents a key index of one's musicality (Sloboda et al. 2005). But the control of the vocal chords and breathing apparatus required to sing in tune is arguably as complex as the technical proficiency required to play a musical instrument, and we should not expect to be able to sing proficiently without some level of training and practice (Welch 2006). The individual differences in the physiological maturation of the vocal apparatus ensures that, in any class of children, there will be some who can hit the right notes, and others who must strive to do so. Unfortunately, this developmental stance is not always appreciated by classroom teachers and peers, and it is frequently heard that adults who self-define as 'tone-deaf' started off as children who were instructed to mime in school assembly. Labelling such children as tone-deaf and requesting that they do not sing has the self-defeating effect of arresting further development of singing, thus ensuring that such individuals will never hit the right notes and will carry the belief into adulthood that they lack musical ability. The damage that such early experiences can have must not be underestimated. Not only do individuals find themselves alienated and often embarrassed from social situations in which singing plays a role, but this often erroneous belief that they are unmusical will exclude them from other potentially fulfilling opportunities to engage with music.

Music-making, *per se*, can really be viewed as the icing on the cake of musicality. The feat of making sense of the music we hear is already a vastly sophisticated behaviour, but the ease with which most of us do this belies the complexity of the process. Music does not exist in the outside world or even at the point at which it hits the eardrums. The patterns of vibrations caused by hitting, plucking and blowing an instrument are made sense of by multiple brain areas across both hemispheres (see for a review of case studies of disordered musical listening Stewart et al. 2003). The frequency and intensity of vibrations are translated into our perceptions of pitch and loudness, while other information in the vibration patterns allows us to deduce which instrument is the source of the sound. But when we listen to an orchestra, or even a string quartet, each instrument generates a different pattern of vibrations at the ear. In a process known as 'streaming', the brain groups the patterns of vibrations from each sound source allowing us to distinguish among interwoven melodic lines. This explains something of how the moment-to-moment patterns of vibrations can be interpreted as musical sound, but to truly make sense of music, we must integrate musical events over time, unify features with those that have gone before and build expectations about those which will follow. The brain is supremely adapted for making links, seeking patterns and creating order from chaos.

Studies with infants reveal that we are born with the ability to extract the rules of music. Just as with language, musical perceptual ability is driven by innately specified brain mechanisms and by the input provided in the environment (Streeter 1976). It is widely known that infants are sensitive to non-native speech sounds that adults simply fail to perceive (Werker and Tees 1984). Similar principles apply within music. Elegant studies using a 'preferential looking paradigm' demonstrated that infants are equally sensitive to deviations in a native and a non-native metre. Metre can be thought of as a grouping unit of pulses. In a waltz, for instance, pulses are grouped in threes: **one**, two three, **one**, two, three; in a march, they are grouped in fours: **one**, two three, four; **one**, two, three, four. By contrast, in much Eastern European folk music, pulses are organised in fives or sevens, giving the music a distinctly irregular feel to Western ears. Hannon and Trehub (2005a, 2005b) played Canadian infants both Western and Balkan rhythms, and introduced a metrical deviation to both. While adults could only detect a deviation to their native metre; children of 6 months and younger were equally sensitive to deviations in both, while infants of 12 months were, like Canadian adults, only sensitive to deviations in their native metre. The older infants had undergone a 'perceptual narrowing', making them more adept at discerning the musical structure of their own environment. However, with repeated exposure to Balkan folk music, the older infants (but not the adults) started to behave similarly to the younger infants, revealing that perceptual abilities are not irreversible and are intimately related to environmental input.

Musical affinities

The affinity humans have for music appears to be universal: no culture in recorded history has been without some form of music which invariably forms a crucial part of rituals and ceremonies. But its survival value is not obvious, leading some to suggest that music is a spandrel: 'Music is auditory cheesecake. It just happens to tickle several important parts of the brain in a highly pleasurable way, as cheesecake tickles the palate' (Pinker 1997, 525). Others disagree and suggest that the pleasure is merely part of music's adaptive role in mother–infant bonding, an extension and development of our language abilities or a product of sexual selection, (see Cross 2007 for a review) depending on which theory one supports.

Regardless of music's evolutionary origins, the power of music to move us is considerable. At one extreme, some listeners experience the phenomenon of 'shivers down the spine', an episode of intense physiological arousal, triggered by a particular piece or passage in a piece. One of the classic brain imaging studies to look at the emotional responses to musical listening was conducted by Blood and Zatorre (2001). These authors used functional magnetic resonance imaging (fMRI) to scan listeners as they experienced a shiver episode. A clever feature of the design was that one person's 'trigger' piece was another person's control piece, since the pieces which trigger shivers tend to be idiosyncratic. When the brain activity involved in listening while experiencing a shiver was compared with the brain activity involved in just listening, activity was seen in the same areas that are involved in other pleasurable activities: sex, eating and the consumption of illicit substances (chiefly cocaine).

What precisely is it about music that we are responding to emotionally? One popular idea, first suggested by Meyer (1956), posits that the building of expectancies during musical listening is key to appreciation and emotion. Through a lifetime of exposure to the music of our culture, we become sensitive to the regularities used by composers of different genres, and we effortlessly and unconsciously predict how the music we are listening to will unfold. The skills of the composer are all-important in crafting our expectancies, and because music is such a multidimensional stimulus, expectancies can be formed at many different levels (pitch and rhythm being two obvious examples).

This multidimensionality may explain why individuals who are compromised with respect to one aspect of music (e.g., pitch), may still derive pleasure from listening to some types of music. Congenital amusia is a disorder in which individuals with normal hearing nevertheless have difficulty in making sense of music (Ayotte, Peretz, and Hyde 2002). The core impairment seems to be in pitch processing (Foxton et al. 2004; Hyde and Peretz 2004), with many of them showing normal processing of rhythmic information. However, although anecdotal reports have focused on those amusics who find music 'aversive', a recent study showed that, within a group of 21 amusics, a subgroup of individuals report levels of engagement with music that are equivalent to those shown by non-amusic controls (McDonald and Stewart 2008). The ability to engage in music and appreciate it despite diminished perception of a key aspect of music is likely to depend on deliberate and conscious exposure to forms of music that involve dimensions in addition to the one that is compromised, facilitating the building of expectancies upon which musical appreciation can be based.

Does practice make perfect?

Turning from the perception and appreciation of music, we now focus on musical performance, which many consider to be the epitome of musical achievement. Two questions to consider are whether neuroscience supports the notion that musical achievement can be explained by notions of giftedness and talent and whether neuroscience, to date, has offered musicians ways of facilitating and optimising their musical learning and performance.

It is a commonly held belief that the ability to play music well requires a special gift or talent. Defining what is meant by the terms 'gifted' or 'talented' is critical to this argument, since a discussion of their contribution to the attainment of musical excellence can otherwise become circular (Question: 'How do you know someone is talented?' Answer: 'Well, just listen to how beautifully they play.'). One approach is offered by Gagné's *Differentiated model of giftedness and talent* (Gagné 1985, 1993, 2000, 2003). At the heart of the model is the distinction between domains of ability (gifts) and fields of performance (talent). Gagné uses the term giftedness to describe individuals in the top 10% of the population for their age who are endowed with natural *potential to achieve* that is distinctly above average in one or more aptitude domains. In this conception, aptitudes are natural abilities that have a genetic origin and that appear and develop more or less spontaneously in every individual. The mix of these aptitudes explains the major proportion of differences between individuals when the surrounding environment and practice are roughly comparable. However, aptitudes do not develop purely by maturation alone; environmental stimulation through practice and learning is also essential (see McPherson and Williamon 2006 for a review).

With these defining features in mind, the scientific evidence for natural gifts leading to specific musical talents has thus far been scant (see Howe, Davidson, and Sloboda 1998). Key to the argument against the existence of musical giftedness is the fact that high levels of musical achievement are rarely preceded by unequivocal signs of musical precocity, and the fact that individuals who are not deemed to be 'gifted' can, nevertheless, attain equal levels of excellence as those who are, given adequate provision of opportunity and encouragement. As a result, scientists have turned to investigations of musicians' early experiences, preferences, opportunities, habits, training and practice to understand the acquisition of performance skills.

Evidence from neuroscience provides key insight into the significance of these factors – particularly training and practice – in the attainment of musical excellence. Several studies have revealed structural differences in the brains of musicians: the corpus callosum (Schlaug et al. 1995) and certain auditory (Schneider et al. 2002) and motor (Amunts et al. 1997) regions have all been shown to be structurally enlarged. However, congruent with the behavioural evidence cited by Howe, Davidson, and Sloboda (1998), the brain imaging data supports the view that deliberate practice is the prime predictor of these changes. The structural alterations seen scale in magnitude with the age at which training commences (Elbert et al. 1995; Hutchinson et al. 2003) and the overall intensity of training over the lifespan (Gaser and Schlaug 2003; Hutchinson et al. 2003; Schneider et al. 2002). Although neuroscientific research may not exactly confirm the time-honoured adage that 'practice makes perfect' (indeed, very little in music can be classified as 'perfect'), it does support the notion that 'practice makes better' and offers clear and observable evidence for how the brain adapts to the demands of extensive training, enabling musical skills to flourish.

Learning in action

Given the pre-eminent role of learning in the development of musical excellence, it becomes pertinent to ask how musicians acquire and integrate the requisite cognitive, motor, perceptual and social skills that enable the most eminent among them to redefine the upper limits of human intellectual and motor achievement.

The work of the first author (L. Stewart) has focused specifically on one aspect of a musician's repertoire: music reading. A longitudinal study conducted with a group of adult non-musicians revealed that after three months of piano lessons, the brains of these individuals showed a different pattern of activation in response to musical notation. Specifically, there were separate learning-related changes for the melodic and rhythm aspects of the notation (Stewart et al. 2003). The changes seen for melody were in the parietal cortex, an area involved in visuospatial processing, while the changes seen for the rhythmic aspects of notation were in early visual cortex, an area associated with performing visual discriminations. The intuitions of performing musicians are that the melodic and rhythmic information contained within single notes are processed simultaneously, while the findings suggest that they depend on different specialisations within the brain. This is true of our visual perception of the world in general: our experience is unified, even though the brain has to combine information processed from different functionally specialised areas.

A particularly interesting aspect of the study was that the changes in the brain's response to notation happened after only a short period of training and that, even at this early stage in learning, the brain had started to process musical notation automatically. The evidence for this was twofold. First, a 'musical-Stroop' task showed that after training, musical novices could not ignore musical notation in the context of a task that required them to make key presses on the basis of numbers. More precisely, a bar of musical notation was presented, on to which numbers between 1 and 5 had been superimposed. Instructed to ignore the musical notation and make key presses on the basis of the numbers alone (1 = thumb, 2 = index finger, etc.), the musical learners, post-training, were affected by the congruence of the number/note pairing. When both the note and the number conveyed the same action, they were faster; when the note and the number conveyed a different action, they were slower. A control group who had received no musical training were insensitive to this note/number pairing (Stewart, Walsh, and Frith 2004). The second line of evidence comes from the neuroimaging data (Stewart et al. 2003). During a condition where participants were instructed to search for a visual target amid bars of musical notation or bars of 'false notation' (visually complex but musically meaningless), they showed activation in the area of parietal cortex that was involved in the explicit melody reading condition. Simply seeing musical notes after a short period of musical training sets in motion a whole string of neural events related to the learned musical responses conveyed by the musical notation. Just like the results of the 'musical Stroop' experiment, these changes in brain activity show that musical training causes notes to acquire a significance that cannot be suppressed.

Expert performance in action

Complementary to the work on musical learning in the beginner is research with musical experts. Given the methodological limitations of collecting data during *actual* performances, where musicians must physically manipulate their instruments, move about on stage and interact with audiences, neuroscientists have turned to examining musicians' *mental* rehearsal and performance processes, asking them to think about specific performance situations and compositions without making corresponding physical movements. Findings reveal that brain areas hypothesised to be involved in the perception and actual performance of music show similar activation patterns during the mere imagery or mental practice of music. In an fMRI study with six adult musicians, Langheim et al. (2002) found that imagined performances of self-chosen musical excerpts activated a network of cortical areas that are supposed to integrate motor and musical-auditory maps in the brain (right inferior frontal gyrus) and musical and motor timing aspects (bilateral lateral cerebellum), as well as spatial features of motor and pitch representations (right superior parietal lobule). In an EEG study, Petsche, Von Stein, and Filz (1996) analysed the pattern of brain activity in one cellist during different experimental tasks. Their results suggest that the supplementary motor area (SMA) is most engaged with mentally playing scales and, less so but still observable, with imagining playing the piece and even with mere listening. Similarly, Halpern (2001) found that the SMA is involved in musical imagery and has suggested that the SMA may provide rehearsal mechanisms such as imaginary humming along with the imagined music.

Nevertheless, such studies have yet to examine one of the key elements of music performance: how, and to what extent, do changes in brain activation correspond to rehearsal, retrieval and expressive mechanisms employed at different points *throughout* a performance? Knowing, for instance, that networks of cortical activation are the same for an entire mental performance as for an entire physical performance is valuable, but in relation to how expert musicians learn and prepare music, the natural next step is to explore how these networks change *within* comparable mental and physical performances. This would divulge significant insight into whether the psycho-physiological processes that enable both are actually the same and, if so, could provide a stronger justification for integrating well-targeted mental rehearsal techniques into the education and training of skilled musicians (e.g., as a preventative measure for musculoskeletal problems resulting from too much physical practice).

However, detecting such changes, especially given that music performance is an inherently physical activity, is no trivial feat. Using current neuroimaging techniques, movement by participants can lead to artefacts in the data, in some cases, rendering them useless. Nevertheless, recent neuroscientific research has employed innovative experimental protocols and data analysis techniques to overcome such methodological obstacles. These have included *in vivo* measurements of musicians playing keyboard instruments (e.g., Parsons et al. 2005). Other studies have systematically extracted hypotheses from behavioural research for systematic testing in the laboratory.

For instance, musicians' apparent ability to transcend the limitations of memory during performance has long since captured the public's imagination. A series of studies by the second author (AW) set out to examine how musicians come to learn and memorise music for performance (Williamon and Valentine 2000, 2002; Williamon, Valentine, and Valentine 2002). Results of this research suggest that musicians, irrespective of their skill level, frequently start and stop their practice on bars in the music that they identify as integral to the music's structure (or form). Moreover, the prevalence of this practice strategy apparently increases as musicians learn a piece, from their initial practice session to a polished performance. This pattern is most pronounced for musicians at the highest levels of skill, and as such, the data appear to confirm some key predictions of psychological theories of expert memory – namely, that exceptional memory relies on the formation and exploitation of highly ordered retrieval structures and that these structures are most stable when rehearsed extensively throughout the learning process (Ericsson and Kintsch 1995).

None the less, these behavioural studies say little about the neural substrates of musical memory, and given that so much behavioural data has confirmed the prevalence of musical structure in musicians' encoding and retrieval processes, a follow-up study using laboratory-based, psycho-physiological measures was designed to investigate whether structurally important bars in a piece (as defined by performers themselves) were indeed integral to the encoding and retrieval of that piece from memory (Williamon and Egner 2004). A recognition memory task was devised that required a group of advanced pianists to identify bars from a piece of music they had recently learned to play from memory. The pianists were asked simply to provide a 'yes/no' response when presented with individual bars on a computer screen – that is, to distinguish bars belonging to the piece from similar bars not belonging to the piece, using a verbal response. Of interest was whether responses to hypothesised 'structural' bars would differ, in terms of response times and event-related potentials (a form of EEG activity), from bars that also belonged to the piece but were presumed to be 'non-structural'. Thus, even though structural and non-structural bars belonged to the same response category in the recognition task (piece versus non-piece), different behavioural and cortical responses to these stimuli were expected. The results confirmed this hypothesis. The recognition of structurally crucial moments in the music was accomplished with greater ease, and they were distinguishable from other segments of the music in terms of the brain activity underlying their retrieval. Although the recognition task involved was itself unmusical, the tested hypothesis was derived from studies that maintained ecological (and musical) validity. The concurrence of behavioural and psycho-physiological results thereby lends weight to the methods used and to the conclusions drawn in the research.

Optimising physical and mental skills for performance

Musicians are not only concerned with how best to rehearse, but how they can achieve an optimal state for performance on stage. Performance anxiety is commonplace among professional musicians and can be career-halting (Fishbein et al. 1988). Thus, research into the emotional as well as technical aspects of musical performance is necessary.

'Zoning In: Motivating the Musical Mind' was a Leverhulme-funded project based at the Royal College of Music from 1999 to 2002. Its aim was to enable music students to improve their performance skills and manage the high levels of stress that often accompany performance situations. Over 150 students at the College worked with a team of scientists and musicians to learn complementary mental and physical training routines drawn from four areas: (1) physical fitness, (2) Alexander technique, (3) neurofeedback and (4) mental skills training.

One of these interventions, neurofeedback, which refers to the monitoring of one's own brain activity with a view to influencing it, delivered marked improvements in the performance ability of the participating students (compared with control and other comparison groups). Moreover, the improvements were highly correlated with their ability to progressively influence neural signals associated with attention and relaxation (Egner and Gruzelier 2001; Gruzelier and Egner 2004). Similar results have subsequently been found with dancers (Raymond et al. 2005). This is an unusual example of a technique being borrowed from neuroscience to provide direct improvements for learners. Throughout the project, the team of collaborating scientists and musicians worked together to shape the delivery of each intervention and develop methods of assessing their effects. As such, one major outcome of the project, in addition to peer-reviewed articles and an edited book, has been a course unit now embedded in the RCM's undergraduate curriculum: *Psychology of performance* provides an introduction to performance enhancement research, along with theoretical seminars and practical training in select 'Zoning In' techniques (see Williamon 2004).

Future directions for neuroscience research and applications in musical education

The neuroscientific study of musical learning and performance is ripe for development, allowing insight into how people acquire and integrate the cognitive, motor, perceptual and social skills that can be at work during listening and performance. Clearly, neuroscience has much to gain from investigations of this distinctive and seemingly universal human behaviour. But what do musicians gain?

Williamon and Thompson (2004) have argued that current scientific research has begun to offer theoretical and pedagogical understanding of practical aspects of music learning and teaching and that the potential for further growth is now greater than ever. Moreover, several scientists have undertaken extensive programs of applied research with the aim of providing advice directly to musicians and their teachers on how to acquire specialised skills more easily and quickly (see Williamon 2004 for a review).

Given that music listening and performance are inescapably cognitive tasks, perhaps no other science promises more to music than cognitive neuroscience. In this paper, we have highlighted significant examples of how this field has already provided insight into processes and mechanisms that underpin musical listening, learning and performance. Nevertheless, there are distinct challenges that interested parties – neuroscientists, performers, music teachers, and so on – must overcome.

First and chief among these is the need to maintain ecological validity of scientific paradigms. We listen to music on the move, at the cinema, in concerts and elsewhere, and on occasion, it can have tremendous influence over us, both motionally and emotionally. Capturing that experience in the laboratory is not easy, and until such a time that laboratories are equipped to allow people to engage fully in *musical* experiences (be they solitary or social), we must endeavour to triangulate findings in the laboratory with those that emerge from behavioural and observational studies.

A second challenge for those interested in exploring the implications of neuroscience for musical education, and vice versa, is to conduct musically meaningful research with rigorous scientific outcomes. Research studies that aim to address musically relevant questions but that are, simply, empirically intractable will not lead to substantive conclusions. Likewise, research that endeavours to elucidate music cognition and perception without any remote connection to what musicians and listeners actually do will offer little by way of real-world application. In order to bridge this gap, research teams should strive to be, by default, interdisciplinary, where both scientists and musicians set the agenda, offer hypotheses, carry out day-to-day investigation and scrutinise results. Only then will the full and mutual benefits of true interdisciplinarity be realised.

References

Amunts, K., G. Schlaug, L. Jancke, H., et al. 1997. Motor cortex and hand motor skills: Structural compliance in the human brain. *Human Brain Map* 5: 206–15.

Ayotte, J., I. Peretz, and K. Hyde. 2002. Congenital amusia: A group study of adults afflicted with a music-specific disorder. *Brain* 125: 238–51.

Blacking, J. 1995. *Music, culture, and experience*. Chicago: University of Chicago Press.

Blood, A.J., and R. Zatorre. 2001. Intensely pleasurable responses to music correlate with activity in brain regions implicated in reward and emotion. *Proceedings of the National Academy of Science USA* 98: 11818–23.

Cross, I. 2007. Music and cognitive evolution. In *Oxford handbook of evolutionary psychology*, ed. R. Dunbar and L. Barrett, 649–67. Oxford: Oxford University Press.

Cuddy, L.L., L.-L. Balkwill, I. Peretz, and R.R. Holden. 2005. Musical difficulties are rare: A study of 'tone deafness' among university students. *Annals of the New York Academy of Science* 1060: 311–24.

Draganski, B., C. Gaser, V. Busch, G. Schuierer, U. Bogdahn, and A. May. 2004. Neuroplasticity: Changes in grey matter induced by training. *Nature* 427, no. 6972: 311–12.

Egner, T., and J.H. Gruzelier. 2001. Learned self-regulation of EEG frequency components affects attention and event-related brain potentials in humans. *Neuroreport* 12: 411–15.

Elbert, T., C. Pantev, C. Wienbruch, B. Rockstroh, and E. Taub. 1995. Increased cortical representation of the fingers of the left hand in string players. *Science* 270: 305–7.

Ericsson, K.A., and W. Kintsch. 1995. Long-term working memory. *Psychological Review* 102: 211–45.

Fishbein, M., S.E. Middelstadt, V. Ottati, S. Strauss, and A. Ellis. 1988. Medical problems among ICSOM musicians: Overview of a national survey. *Medical Problems of Performing Artists* 3: 1–8.

Foxton, J.M., J.L. Dean, R. Gee, I. Peretz, and T.D. Griffiths. 2004. Characterization of deficits in pitch perception underlying 'tone deafness'. *Brain* 127: 801–10.

Gagné, F. 1985. Giftedness and talent: Reexamining a reexamination of the definitions. *Gifted Child Quarterly* 29: 103–12.

———. 1993. Constructs and models pertaining to exceptional human abilities. In *International handbook of research and development of giftedness and talent*, ed. A. Heller, F.J. Monks, and A.H Passow, 69–87. New York: Pergamon.

———. 2000. Understanding the complex choreography of talent development through DMGT-based analysis. In *International handbook of giftedness and talent*. 2nd edn, ed. K.A. Heller, F.J. Monks, R.J. Sternberg, and R.F. Subotnik, 67–79. New York: Elsevier.

———. 2003. Transforming gifts into talents: The DMGT as a developmental theory. In *Handbook of gifted education*. 3rd edn, ed. N. Colangelo and G.A. Davis, 60–74. Boston, MA: Allyn & Bacon.

Gaser, C., and G. Schlaug. 2003. Brain structures differ between musicians and non-musicians. *Journal of Neuroscience* 23: 9240–5.

Gruzelier, J.H., and T. Egner. 2004. Physiological self-regulation: Biofeedback and neurofeedback. In *Musical excellence*, ed. A. Williamon, 197–219. Oxford: Oxford University Press.

Halpern, A.R. 2001. Cerebral substrates of musical imagery. In *The biological foundations of music*, ed. J. Zatorre, and I. Peretz, 179–92. *Annals of the New York Academy of Science* 930.

Hannon, E.E., and S.E. Trehub. 2005a. Metrical categories in infancy and adulthood. *Psychological Science* 16: 48–55.

———. 2005b. Tuning into musical rhythms: Infants learn more readily than adults. *Proceedings of the National Academy of Science USA* 102: 12639–43.

Howe, M.J.A., J.W. Davidson, and J.A. Sloboda. 1998. Innate talents: Reality or myth? *Behavioural and Brain Sciences* 21: 399–442.

Hutchinson, S., L.H. Lee, N. Gaab, and G. Schlaug. 2003. Cerebellar volume of musicians. *Cerebral Cortex* 13: 943–9.

Hyde, K., and I. Peretz. 2004. Brains that are out of tune but in time. *Psychological Science* 15: 356–60.

Langheim, F.J.P., J.H. Callicott, V.S. Mattay, J.H. Duyn, and D.R. Weinberger. 2002. Cortical systems associated with covert music rehearsal. *NeuroImage* 16: 901–8.

Maguire, E.A., D.G. Gadian, I.S. Johnsrude, C.D. Good, J. Ashburner, R.S. Frackowiak, and C.D. Frith. 2000. Navigation-related structural change in the hippocampi of taxi drivers. *Proceedings of the National Academy of Science USA* 97, no. 8: 4398–403.

McDonald, C., and L. Stewart. 2008. Uses and functions of music in congenital amusia. *Music Perception* 25: 345–55.

McPherson, G.E., and A. Williamon. 2006. Giftedness and talent. In *The child as musician*, ed. G.E. McPherson, 239–56. Oxford: Oxford University Press.

Meyer, L. 1956. *Emotions and meaning in music*. Chicago: University of Chicago Press.

Parsons, L.M., J. Sergent, D.A. Hodges, and P.T. Fox. 2005. The brain basis of piano performance. *Neuropsychologia* 43: 199–215.

Peretz, I., A.-S. Champod, and K.L. Hyde. 2003. Varieties of musical disorders: The Montreal battery of evaluation of amusia. *Annals of the New York Academy of Science* 999: 58.

Petsche, H., A. Von Stein, and O. Filz. 1996. EEG aspects of mentally playing an instrument. *Cognitive Brain Research* 3: 115–23.

Pinker, S. 1997. *How the mind works*. New York: W.W. Norton.

Raymond, J., I. Sajid, L. Parkinson, and J.H. Gruzelier. 2005. The beneficial effects of alpha/theta and heart rate variability training on dance performance. *Applied Psychophysiology and Biofeedback* 30: 65–73.

Schlaug, G., L. Jancke, Y. Huang, J.F. Staiger, and H. Steinmetz. 1995. Increased corpus callosum size in musicians. *Neuropsychologia* 33, no. 8: 1047–55.

Schneider, P., M. Scherg, H.G. Dosch, H.J. Specht, A. Gutschalk, and A. Rupp. 2002. Morphology of Heschl's gyrus reflects enhanced activation in the auditory cortex of musicians. *Nature Neuroscience* 5: 688–94.

Sloboda, J.A., K.J. Wise, and I. Peretz. 2005. Quantifying tone deafness in the general population. *Annals of New York Academy of Science* 1060: 255–61.

Stewart, L., R. Henson, K. Kampe, V. Walsh, R. Turner, and U. Frith. 2003. Brain changes after learning to read and play music. *NeuroImage* 20: 71–83.

Stewart, L., V. Walsh, and U. Frith. 2004. Reading music modifies spatial mapping in pianists. *Perception and Psychophysics* 66: 183–95.

Stewart, L., K. Von Kriegstein, J.D. Warren, and T. Griffiths. 2006. Disorders of musical listening. *Brain* 129: 2533–53.

Streeter, L.A. 1976. Language perception of 2-month-old infants shows effects of both innate mechanisms and experience. *Nature* 259: 39–41.

Welch, G. 2006. Singing and vocal development. In *The child as musician*, ed. G.E. McPherson, 311–29. Oxford: Oxford University Press.

Werker, J.F., and R.C. Tees. 1984. Cross-language speech perception: Evidence for perceptual reorganisation during the first year of life. *Infant Behaviour and Development* 7: 49–63.

Williamon, A., ed. 2004. *Musical excellence*. Oxford: Oxford University Press.

Williamon, A., and T. Egner. 2004. Memory structures for encoding and retrieving a piece of music: An ERP investigation. *Cognitive Brain Research* 22: 36–44.

Williamon, A., and S. Thompson. 2004. Psychology and the music practitioner. In *The music practitioner: Research for the music performer, teacher, and listener*, ed. J.W. Davidson, 9–26. Aldershot: Ashgate.

Williamon, A., and E. Valentine. 2000. Quantity and quality of musical practice as predictors of performance quality. *British Journal of Psychology* 91: 353–76.

———. 2002. The role of retrieval structures in memorizing music. *Cognitive Psychology* 44: 1–32.

Williamon, A., E. Valentine, and J. Valentine. 2002. Shifting the focus of attention between levels of musical structure. *European Journal of Cognitive Psychology* 14: 493–520.

Wilson, G.D. 1994. *Psychology for performing artists: Butterflies and bouquets*. London: Jessica Kingsley.

Co-constructing an understanding of creativity in drama education that draws on neuropsychological concepts

Paul A. Howard-Jones[a], M. Winfield[b] and G. Crimmins[b]

[a]Graduate School of Education, University of Bristol, UK; [b]Cardiff School of Education, University of Wales Institute Cardiff, UK

Background: Neuroscience is unlikely to produce findings for immediate application in the classroom. The educational significance and practical implications of knowledge about mind and brain inevitably require some level of interpretation, yet the multiplying examples of unscientific 'brain-based' educational concepts suggest this process of interpretation is potentially problematic. Research is needed into the most appropriate ways of developing such concepts.
Purpose: This paper reports on an attempt to develop a process of 'co-construction' of pedagogical concepts, enriched by insights about the brain and the mind, with a group of trainee teachers led by a team with both educational and scientific expertise.
Sample, design and methods: A research team consisting of two teacher trainers and a psychologist followed an action research spiral involving 16 trainee teachers who explored their own creativity, and the psychology and cognitive neuroscience of creativity in seminars, discussions and practical workshops, with the pedagogical aim of developing their own reflective capability.
Results: Outcomes illustrated both dangers and opportunities associated with developing concepts bridging neuroscience and education. Trainees' understanding developed in stages that might broadly be described as initial enchantment, mythologising, disenchantment, an increased focus on metacognition and, finally, a demonstrable ability to reflect on their own classroom practice with a heightened sensitivity to issues of underlying cognitive processes.
Conclusions: The type of 'co-construction' process reported here may help reduce some of the more popular and problematic misconceptions that arise when developing pedagogical concepts involving the brain and mind. Further research is needed to assess impact of such concepts upon practice.

Introduction

An important area of challenge for the new interdisciplinary area of neuroscience and education is the culturing of pedagogical ideas that appropriately combine educational knowledge with concepts about the brain and the mind. History has already demonstrated how this can happen in a variety of unsatisfactory and often unscientific ways (see Geake, in this issue). As well as the practical usefulness of a pedagogic concept, the validity of any purported scientific basis for its validity is also an important issue, not least because many teachers would like to know not just what works, but why and how (Pickering and

Howard-Jones 2007). This understanding of underlying processes may also contribute to more effective implementation and evaluation. However, the production of credible concepts that span neuroscience and education may rely upon the development of improved communication and language, and the emergence of a two-way dialogue rather than a one-way transfer (Geake 2004). In the project described here, a process of co-construction is pursued by two educators (teacher trainers) and a psychologist with some educational and neuroscientific experience. We report upon efforts to collaborate within one particular context of teacher training, but it is hoped that the insights regarding the process of co-construction may be helpful in developing similar projects in other areas of education.

The chosen context for this study was the fostering of creativity in drama education. The potential complexity and diversity of creative processes made this a somewhat daunting context to work in. However, there is an increasing interest in creativity in the curriculum and a surprising lack of guidance available for trainee teachers in the fostering of creativity, especially in the field of drama education. It was this paucity of current research and understanding that provided the chief motivation for the project reported here which, in pedagogical terms, aimed to develop the reflective capability of trainee drama teachers in regard to the fostering of creativity, through a better awareness of the underlying cognitive and neurocognitive processes involved. Such an aim attends to the calls of those such as Chappell (2007), who has also highlighted the need within teacher training for an increased emphasis upon reflective practice in teaching for creativity. It should be noted, however, that the team did not intend to produce a pedagogical approach based solely upon scientific findings. The inadequacy of neuroscience (including cognitive neuroscience) to provide specific instructions for improving learning has been explored by a number of writers (e.g., Schumacher 2007; Davis 2004) and the team made several excursions during seminars to illustrate the limitations of scientific knowledge within education, when such knowledge is isolated from insights arising from other perspectives. Rather, the approach was to encourage trainees to broaden their reflections upon learning by *including* psycho-biological perspectives, and to provide them with a set of theoretical tools drawing on scientific insights that could be judiciously integrated with their own experience and those educational concepts they had already developed as part of their training.

Questions about the processes by which teachers and trainee teachers might successfully integrate their existing pedagogical knowledge and experience arose during efforts to pursue a wider multi-perspective cycle of research activity involving biological, social and experiential approaches to investigate creativity. This paper focuses only on this issue of developing practical and credible pedagogical concepts, but the wider cycle is reproduced in Figure 1, in order to illustrate the broader research contexts in which the study was undertaken. As part of the wider investigative effort, students attending the same BA course in Drama Education as our present participants had already been involved with a functional magnetic resonance imaging (fMRI) study of a strategy intended to foster creativity (Howard-Jones et al. 2005). (However, none of the trainees participating here had participated themselves in the fMRI study, or received any specialist knowledge of psychology or cognitive neuroscience as part of their under-graduate experience.) This fMRI study had focused upon 'random strategies' – i.e., strategies that require the incorporation of items into a creative outcome that are unrelated to each other and/or any context of the brief. As confirmed by the study, such strategies generally improve the perceived creativity of outcomes, but the fMRI results also showed increases in activity associated with creative effort. This supported the notion

Figure 1. The work reported here is a part of a wider cycle of research activity aimed at increasing understanding about creativity, involving experimentation and more interpretative approaches. The cycle began by consulting with teachers and teacher trainers (top left) to help formulate hypotheses that might be tested using neuroscientific techniques such as functional magnetic resonance imaging (fMRI) (top right). Experiential investigations (bottom right) then examined issues from an 'insider' viewpoint, using theatrical workshops to explore aspects that scientific investigations find typically problematic, such as those associated with free-will and autonomy. Finally (bottom left), the findings from both the 'outsider' scientific studies and the 'insider' experiential investigations were taken forward to the present study, allowing practitioners, with expert support, to take ownership over findings in terms of their educational significance, using these and other findings to co-construct concepts that can support improved reflective practice. Such interdisciplinary dialogue may give rise to further potential research questions.

that the strategies encourage increased processing of a type associated with creative thought, rather than providing an effortless cognitive short-cut to improved ratings. By suggesting they encourage rehearsal of cognitive processes that we might call creative, the results support the likelihood of their being longer-term benefits to the learner. So, this fMRI study produced a finding that might be relevant to practice, but issues quickly arose when we considered how such a finding should be communicated back to educators. First, any individual scientific finding about creativity resides in the context of a larger body of knowledge from psychology and cognitive neuroscience and needs to be understood within that context. For example, without reference to related cognitive models, isolated biological images of blood flow in the brain may be distracting but have little to offer education (Bruer 1997). It was clear that the 'translation' of neuroscientific understanding to the classroom would be fraught with dangers of unscientific interpretation and/or departure from a grounded educational understanding. Building any useful conceptual bridge that spans neuroscience and education would require communication of broader issues and concepts, and co-construction of understanding by those with expertise on both sides. Therefore, in addition to the pedagogical aim identified above, the research aim of the project was to provide an improved understanding of this process of co-construction, since this might be helpful to any future ventures integrating neuroscience and education.

Method

The research team consisted of two teacher trainers and the neuroeducational researcher who directed the original fMRI study. The methods used to communicate concepts and the details of the content covered in sessions was negotiated between members of the research team and informed by the responses of the trainees as the project progressed. In terms of content, note was made of what trainees found useful in terms of understanding their own and their pupils' experiences and learning. In terms of developing communication methods, the research team took particular note of the appropriateness, relevance and validity of the ideas expressed by trainees during sessions.

Sixteen trainee teachers, in the second year of their training, voluntarily took part in what was advertised as a short program of seminars and activity-based workshops exploring concepts about creativity. An action research spiral (Elliot 1991) was followed by the researchers (Figure 2) consisting of an initial meeting of the research team and initial discussion with the trainee teachers, followed by three cycles of research meeting, seminar, workshop and student discussion, ending in a final meeting of the team to reflect upon the project as a whole. Workshops, seminars and trainee discussions were video taped, with informed written permission from the participants. After each of these events, an analysis of the video data was used as a basis for discussions during subsequent research team meetings that deliberated upon progress and revised future plans (see Figure 2). An audio recording was made of these research team meetings and this was transcribed to help track the issues raised and decisions made.

Results and analysis

The processes by which pedagogical concepts were constructed are now reported upon in the chronological order in which they occurred, beginning with data arising from the preliminary discussion with the students, followed by each of the three cycles of activity in turn.

Initial discussion with trainees about how to foster their pupils' creativity

Before introducing any new concepts, we had an initial discussion with the trainees that provided some sense of baseline regarding existing ideas about creativity. As observed by Hayes (2004), although the term 'creativity' is frequently used, its direct definition remains problematic, with recent attempts emphasising the role of factors beyond the level of the individual, and issues of ethics and morality (e.g., Craft 2000, 2006). In the initial discussions, the team drew on a simple definition of creativity as the type of imaginative thinking that produces an outcome possessing some level of originality, as well as some

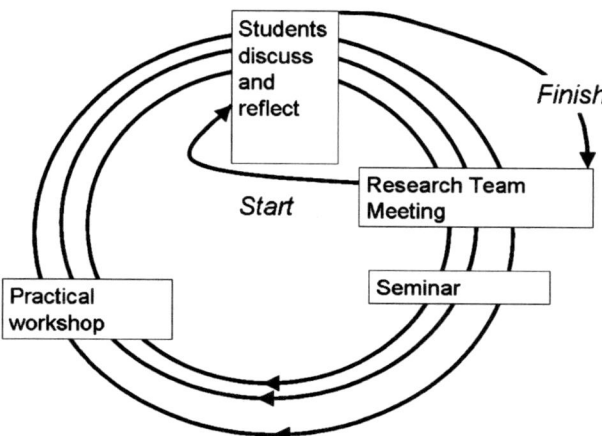

Figure 2. The action research spiral followed by the researchers. After an initial meeting of the research team and discussion with the student participants (trainee teachers), there were three cycles of research meeting, seminar, workshop and discussion with participants, ending with a final meeting of the research team to reflect upon the project as a whole.

sense of value (NACCCE 1999). Trainees felt comfortable with this definition and expressed strong personal convictions about the importance of creativity, a capability that enriched many parts of their lives and was especially appreciated in drama education. Many had chosen to become drama teachers because, as pupils themselves, they had discovered drama was a subject area that embraced creativity. However, creativity was generally seen as a spontaneous process mostly beyond influence and that should simply be allowed to flourish:

> Kids they just – they draw so many things from so many places, and they can bring it all together and they can – and there's your creativity – you can't teach it.

Trainees generally emphasised a 'hands-off' notion of creativity as a type of thinking that appeared in the absence of poor teaching rather than resulting from good teaching. This was evident in the frequent use of phrases such as 'you're allowing them to be creative'.

First cycle

The team agreed that the first priority would be to present a simple cognitive model of creativity. The model used was originally developed to support the teaching of design (Howard-Jones 2002) and describes creative cognition as involving two modes of thinking: generative (G) and analytical (A). The model emphasises the difference between thought processes we use to critically evaluate an outcome and those we use to generate it in the first place, the latter requiring access to concepts that are more remotely associated with the matter at hand. When engaged in analytical thinking, an individual is expected to be focused and to constrain their attention upon the analysis. However, when accessing remote associates, there is benefit from being less focused and allowing attention to drift towards concepts that have not previously been directly associated with the problem. Analytical thinking can also be useful elsewhere in the creative process, such as when researching a topic or context before generating any ideas. Creativity, then, may be characterised by an ability to move from one mode of thought to the other without difficulty. The existence of two distinct modes of thinking is not a new one, but builds on the ideas of Ernst Kris (1952), Wundt (1896) and Werner (1948).

After being introduced to this model of creative cognition, trainees were presented with research illustrating how the conditions for supporting analytical and generative thinking can be quite different. They were reminded how our analytic abilities can often be supported by being encouraged to remain focused, being offered some monetary reward for our performance or by the mild stress of knowing we may be evaluated and assessed. Generative ability, on the other hand, can benefit from changes in context (Howard-Jones and Murray 2003), tasks that require divergent semantic association (Howard-Jones et al. 2005), intrinsic motivations such as fascination and curiosity (Cooper and Jayatilaka 2006) and relaxation (Forgays and Forgays 1992). Production of a single creative idea can require alternation between a focused analytical state when exploring what is known about an issue, a generative state when finding associations beyond the context of the issue itself and a return to the analytical state to assess the value of what has been generated. However, even in the production of a very short story, more complex trajectories between these two modes of thinking can be assumed.

To understand how the creativity of pupils can be directly influenced by a teacher, trainees were introduced to 'random strategies' that require the making of links between elements chosen with some degree of randomness. In the fMRI study discussed in the Introduction to this paper, the neural correlates of creativity in a storytelling task were identified by comparing brain activity when trainees were trying to be creative and to be uncreative as they produced their story (Howard-Jones et al. 2005). Participants had to include a different set of three words for each story. The activity in some areas associated with this creative effort increased further when the words were chosen with some degree of randomness and thus were unrelated to one another. (The creativity of such stories, as assessed by an independent panel of judges, also increased as expected.) The chief area in which correlates of creative effort increased when using this strategy was the right medial gyrus – an area associated with higher-level conscious control, presumably due to increased amounts of filtering out of inappropriate combinations of ideas. So, although the strategy encouraged greater generation of ideas, it may also have required increased amounts of conscious analysis and effort.

In the discussion that followed the seminar, considerable enthusiasm was expressed for using what we know about the brain and mind to enrich pedagogy. Much of the dialogue focused on the fMRI study. The power of brain-imaging to engage interest is well known and research has shown that it stimulates a sense of objective evidence and a 'physicalisation' of concepts of the mind (Cohn 2004). There are attendant dangers in this interest, such as it encouraging notions of static brain states characterised by activity that is restricted to a few limited areas. However, as observed here, it can help 'concretise' psychological concepts that might otherwise remain too abstract to be taken up by non-specialists. Trainees were keen to find real-world analogies with the fMRI experimental task and resonances with their own experience. A trainee reported how she had recently asked every pupil in her class to construct a story around any two of four items: a map, a set of car keys, a ballet shoe and a bottle. Two of these items – i.e., the map and car keys – seemed more obviously related and she noticed the effect on the pupils' creativity:

> the majority of people in the class chose the map and the keys and there were just different variations of car crashes and that was pretty much all they came up with, and the bottle and the ballet shoe – that really worked a lot more creatively.

These observations were, at first, simple behavioural cause–effect links, without any great reference to underlying cognitive processes, and echoing some of the ideas raised in the initial discussion. For example, the trainees, again, seemed to refer to creativity as a spontaneous process, but now as one which required the right level of constraint – not so constrained that it cannot flourish, but requiring enough guidance to provide reassurance. Such ideas have been expressed in studies of creativity in dance education, as a balance between control and freedom (Chappell 2007). It appeared that the trainees' ideas about creativity were becoming more sophisticated, as they suggested that their own creativity sometimes depended on the right level of constraint being provided by their tutor. One trainee reflected upon how she would have felt when performing a particular exercise with such guidance:

> I would have found it quite overwhelming, and I think I would have felt the need to impose guidelines upon myself, but if it's too constrained, then it stifles the creativity and you just don't have the kind of scope required for the kind of work and outcome you want to have.

The idea arose that individual differences existed among learners as to the level of constraint they needed, and this was not necessarily related to academic ability:

> We had a group of super-intelligent girls who sat there for 40 minutes really mulling it over and one of the boys just said to them, 'er ... why don't you do the title "the day I went mad with a spade"?', and they said *'that's it!'* and started writing.

The team suggested that perhaps these girls had been too analytical in their approach and become fixated. Fixation, when one idea or set of ideas becomes overly dominant, had been discussed in the seminar. This prompted the trainees to consider how thinking about creativity in cognitive terms might call into question some aspects of accepted practice, such as target setting and indicating learning outcomes at the beginning of a lesson:

> if you're telling them that at the end of the lesson they're going to be doing a performance, then straight away they're not in generative mode anymore.

As the trainees began to focus more upon underlying cognition, one voiced a realisation that such reflection could radically change their perceptions and their strategy:

> as soon as you build an understanding of how people work, and why they work like that, then you don't necessarily see someone's behaviour in the same way.

A practical workshop followed these discussions. This was aimed at providing trainees with experiences that could later, with support, be linked to some of the scientific concepts of mind and brain they had been introduced to. The workshop included an attempt at identifying what is creative by considering what is perceived as uncreative. Repetition, lack of originality and a tendency towards 'what is obvious' were characteristics that were deemed uncreative. Trainees engaged very actively in this discussion, in contrast to their participation in the next activity, 'Babble', which was a verbal improvisation exercise invented by the team. In 'Babble', students were invited to improvise dialogue by building incrementally from speech-like sounds, through unrelated words to snippets of sentences until they developed a conversation. The team had intended the trainees to engage with the exercise as a form of purposeful play, but the students took up suggestive cues and avoided deviating from them, apparently feeling more comfortable with the type of 'tight apprenticeship' model of learning described by Chappell (2006). However, the team's lack of success in engaging them with this exercise also provided a useful topic for later discussion. It was introduced with few 'rules' and without any physical or imaginative warm-up activity. The subsequent parts of the workshop were more successful. 'Ever-evolving statue' was a familiar physical improvisation exercise in which trainees were required to create physical postures in relation to one another's body positions and shapes. This built from working in pairs to fours to groups of eight. Postures relating to character or narrative development were discouraged in favour of kinaesthetically imaginative interaction. This exercise encouraged trainees to make links echoing the fMRI study, essentially making connections between disparate elements. A 'group morphing' activity provided a movement equivalent of this exercise, and an object improvisation exercise provided another such potential cross-reference between science and experience.

This workshop provided common foci for first reflections upon how ideas emerge. The research team noted the likely importance in developing the trainees' understanding of being able to identify transitions between G and A modes of thinking. So, after the workshop, trainees were asked to produce a line graph indicating where they had been along the G/A continuum at various points in the workshop. Outcomes were very varied but the process prompted trainees to begin reflecting upon their own creative cognitive processes:

> in the last task, you were able to be very, like ... um, generative in the process of creating. And then ... because we were in a group and we knew we had to perform ... we had to bring it back and be, like, analytical ... so my last line is going up and down. We did go back and look at what we were doing ... [laughter], but obviously not enough!

Trainees discussed the ease with which thinking can tend to the obvious, and how it feels when the obvious option is made less available. For example, trainees commented that the items they had selected themselves appeared to them already connected, and they had often begun making a story with them at once. When trainees were required to improvise by linking together unrelated objects selected by the research team, the task became more challenging and difficult, possibly reflecting the additional frontal medial activity observed in the fMRI study of semantic divergence:

> I felt really limited by the fact that you'd given us objects and the fact that we couldn't choose our own … I felt really like I'd hit a wall and was going to have to really think about how I was going to move on.

The trainees identified that a lack of warm-up had contributed to the first ('Babble') exercise going astray, suggesting they needed a way of clearing away some of the unwanted foci of the day to make space for new ideas. There was a sense that everyone had been too willing to focus on the smallest suggestion of a context – a party – and become fixated on it. The trainees then became excited by the importance of relaxation and the generative state, and also discussed how planning one's actions can sometimes diminish generation of ideas. This gave rise to the idea that planning, in which one sets out the stages by which one will achieve a goal, can encourage a particular mind set that discourages generation of new directions and ideas. The trainees appeared comfortable classifying *tasks* as being creative or uncreative and seemed to avoid considering whether they supported the type of thinking required in a particular context. For example, one trainee had begun believing that planning always diminished creativity and the inclusion of randomness always increased it:

> I've got it into my head now that to be uncreative you plan and stuff – so now I think that the last improvisation we did was completely uncreative because I planned it! Because we discussed it as a group and I don't know, now, I'm all confused … I think that the last task was more random … you gave us lots of randomness.

The team gave examples of how different levels of planning can be good or bad for creativity depending on aspects of the situation such as the individuals involved and the types of cognition one might wish to encourage at a particular stage in a creative process. The generative part of creativity had been the main focus of discussion but the team felt it was important to remind them that analysis was also needed. The creative process, as described by Wallas (1926), was presented as a shift from analytical to generative and back to analytical.

Second cycle

There was a clear tendency emerging for trainees to make short cuts from strategies to outcomes without consideration of underlying cognitive processes and context. We needed to diminish the temptation to classify strategies as creative or uncreative, and to encourage the trainees to think more about the appropriateness of strategies in terms of the cognitive processes and whether, in terms of the context, these might be helpful in progress towards creative targets. It was clear that some of the students felt daunted by this task. The team identified the abstract nature of the cognitive concepts involved as a potential challenge for some. We wanted to make the cognitive model of creativity we had been referring to more concrete for the trainees. The trainees had been notably fascinated by a neuroscientific case study mentioned previously by the team, so it was decided to detail two such studies in the next seminar to illustrate extreme examples of the two modes of thinking. This was felt appropriate in the context of training teachers, but the use of such case studies with children would clearly raise some ethical issues. The team felt that classroom discussions about disorders of the mind might easily lead to misconceptions that could distress/confuse some pupils, if teachers leading the discussions were not versed in the necessary expertise.

In the next seminar, the trainees were introduced to a part of the brain called the cingulate cortex – an island of the cortex below the external surface of the brain. The front (anterior) part of this region shares a controlling function with the frontal lobes and is associated with executive attention – the cognitive mechanism by which we control the focus of our attention (Gehring and Knight 2000). Hyperactivity in this area has been associated with Obsessive Compulsive Disorder (OCD) and the associated preoccupation of sufferers with correcting perceived mistakes (Fitzgerald et al. 2005). The trainees were played an interview with a sufferer of OCD, who described her ritualistic repetitive routines. It was discussed how this type of rehearsal resembled the analytical and evaluative rehearsal processes used to hone a piece of creative work, but taken to an obsessive and very uncreative extreme. It was as if sufferers of OCD are caught in an analytical mode of thinking. In contrast, the team then presented a case of compulsive creativity (Lythgoe et al. 2005). The trainees were told that Tommy was a 51-year-old builder with no previous interest in the arts, who suffered a subarachnoid haemorrhage – a bleeding in the space around the front of the brain – resulting in frontal dysfunction. In the weeks following his injury, Tommy became a prolific artist. He first began filling notebooks with poetry, then began sketching and in the following months produced large-scale drawings on the walls of his house, sometimes filling whole rooms. His artistry continues to this day and has become more developed. Tommy cannot stop generating material, often only sleeping 2–3 hours a night between days filled with sculpting and painting. He shows verbal disinhibition, albeit creatively, by constantly talking in rhyming couplets and there are some signs of impaired executive function. Trainees discussed how Tommy appeared to be caught in a generative mode of thinking. Trainees listened to an interview with Tommy who explained what his world was like and they read a poem, 'Brain explorer – it's for you', that he had written for the author of his case study. The team hoped that listening to the voices of those suffering from very generative or analytical mental states would help characterise these modes of thinking more clearly for the trainees and support them in monitoring their own modes of thinking.

In the improvisational exercises that followed, trainees were occasionally interrupted and asked to hold up G or A cards to indicate their current mode of thinking. The first two exercise was 'talk for a minute', in which they had to speak without pause or hesitation on a topic chosen for them. That was followed by a 'delayed copying' exercise in which students had to continuously reproduce not the movement just made by the leader, but the movement previous to it. Trainees almost always held up the generative card when interrupted during the first exercise and the analytical card during the second. When talking-for-a-minute, trainees generated ideas with little time to reflect and reject unsatisfactory elements. When copying movements, trainees focused on a very specific routine, analysed what they saw and rehearsed this mentally before reproducing it. A more complex task followed called 'story in the round', in which trainees sat in a circle and, when asked, had to continue the story their neighbour had been telling. This produced a spread of A's and G's, which trainees explained in terms of individual differences in approach, but also according to where in their own creative process they were when asked to report. Trainees often held up a 'G' when generating links between their ideas and the story their neighbour was telling, or produced an 'A' when evaluating possible stories or those they were hearing. 'Tag improvisation', in which trainees had to step into an improvisation and take over from another performer, also provided an example of this complexity.

Trainees were then asked to produce a piece of movement using the textures and sounds they had encountered during an imaginary journey into a magic wardrobe. Researchers observed and interrupted when they identified points of transition, asking whether trainees were aware that a transition had occurred and whether they could explain why it had happened. Although some trainees had been initially unaware that transitions were even occurring, they quickly began recognising them. They often chose to explain them in terms of a need to move from one mode of thinking to the other. Transitions to rehearsal were often justified in terms of a need to evaluate and hone what had been generated, and thus any attempt to run through the work in progress was usually seen as a return to a more analytical thinking mode. This was something of a turning-point in the project, and the subsequent discussion developed a new richness and depth in terms of the trainees thinking about their experiences in the workshop itself and also their teaching.

Trainees began talking in reflective and often emotional terms about generating and analysing material. Generative processes were described in both positive and negative terms, as highly pleasurable but also slightly frightening. One trainee also described how analytical rehearsal, as in OCD, can become an unhelpful response to anxiety – i.e., the apprehension of having to generate ideas:

> when I'm creating work I feel like I have to keep going back, and like you said: 'what would happen if I didn't go back?' I don't know, but that's what I'm too afraid to find out, I couldn't just keep on creating.

The generative process was described as 'scary', 'like a void' but also as a 'delight', with the workshop reminding trainees how much they enjoyed being generative. Again, the spontaneous nature of creativity that had been mentioned in the earliest session arose, but this time spontaneity was assigned to a particular part of creativity: the ability to generate. The trainees had observed how young children can be highly generative in their thinking, although often less developed in their ability to critically rehearse their ideas. Adults, on the other hand, often find it difficult to maintain such effortless generation of ideas, needing instead to pause, analyse and refine meaning:

> when you told us to talk for a minute, I think the poem [by Tommy] is what we find so hard to do. Like in the poem where there's no links, you said to us don't worry about the links, but automatically everybody tried to make a story even when you'd told us that we didn't need to.

Metacognitive awareness, to the extent of regulating as well as monitoring cognitive processes, became evident:

> I started off by being analytical, thinking: 'What am I expected to get out of it? What am I supposed to be doing with this visualisation?' And then I just thought, 'No, right, cut that off, just leave it, let it go, and just made myself switch off that'.

Interjection by the research team during salient moments of transition not only raised awareness of cognition, but also appeared to encourage self-regulation:

> I knew I was trying to change it, and I knew you'd go, 'Why?'...but then I'd go, 'Oh, I'm being too analytical, let's just change it, let's just go with something different and not keep knocking our head against this brick wall'.

Third cycle

At the next research meeting, the team selected two pieces of footage from previous workshops that would be suitable for analysis with the trainees at the next seminar. At this final seminar, the team first showed footage of the failed 'Babble' exercise from the first workshop, and some excerpts from the discussion with trainees that had followed it. In reflecting upon the outcomes of the exercise, trainees watched themselves improvising on film and afterwards discussed the considerable repetition within and between individuals, the regular occurrences of blocking during the improvised dialogue and a tendency towards fixating upon cues from the team, and noted the feelings of discomfort and obligation that had been discussed afterwards. In understanding why the exercise had not succeeded in generating ideas, discussion centred on feelings of anxiety about not knowing what was required and the lack of a relaxation exercises. Additionally, the preceding tasks had been very analytical in their goal, including analysis of the term 'uncreative' and writing an 'uncreative' story which most students achieved by the self-imposition of constrained boundaries and use of frequent repetition. This may have impacted on generative tendencies in the subsequent exercise, a type of transfer that has been observed elsewhere (e.g., Howard-Jones, Taylor, and Sutton 2002). It was discussed whether seeing a member of the team carry out the task first would have helped. This gave rise to a discussion about mirror neurons which, it has been speculated, may provide a basis for the embodiment of cognition and even the unconscious communication of mental states (Rizzolatti et al. 2002).

Options were considered regarding what might have been done after the failure of this exercise. The trainees were asked: 'Should we have stopped and evaluated what had gone wrong?' 'Should we have gone into some relaxation exercises?' 'Should we have just ploughed on to the next exercise?' It was agreed that an evaluative exercise would probably have further entrenched everyone in an analytical mode of thinking. Recalling the effects of relaxation on free association (Forgays and Forgays 1992), there appeared a clear case for relaxation exercises. Continuing directly on to the next exercise (which is what actually happened) was the more uncertain course which, as it turned out, worked well. The trainees were then asked to consider why it might be that this subsequent exercise (object improvisation) did work better. Three issues emerged from the discussion. First, it was a familiar exercise and the trainees immediately felt more relaxed. Second, the task required links to be made between objects that the trainees had not selected themselves. Third, the trainees felt they had time within the exercise to produce ideas which, as discussed above, may be needed in order to select appropriate links between elements that are disparate. So, the trainees were asked: 'If this was your class and you found one group was staying focused on the brief, asking a lot of questions about boundaries and unable to generate ideas beyond the obvious, what would you do?' Alternatively: 'If another group rushed straight into the improvisation and were generating a lot of incoherent ideas that were not being developed appropriately, what would you do?' In this way, the trainees were encouraged to start thinking about their effect, as teachers, on the creative cognitive processes of their pupils.

After this session on analysis, the trainees were 'hot-seated' about reflections on their own practice. Volunteers took turns to sit in front of the group and recall specific instances in their own practice for discussion and analysis by the group, which now often included reference to their pupils' modes of thinking. For example, it was discussed how questions about procedure and process often reflected an insecure adherence to analytical processes, and how the confidence to create was often accompanied by a diminishment in questioning the teacher. Lower-ability groups often suffer from this lack of confidence, and another trainee drew attention to how a teacher's response to questioning can also be used to orientate pupils' modes of thinking. This trainee described how she used 'teacher-in-role' and then prompted pupils' interpretations. Questions from the class about whether their idea was correct were deflected by the response 'it's whatever you think it is', leaving the arena open for other pupils while legitimising all suggestions as valid self-generated ideas. At first it was the louder children who were questioning her for the right answer, but then, when it was clear that none existed, the quieter children came forward with their ideas. The use of 'teacher-in-role' prompted many other accounts of how pupils can be directed towards a particular mind state through imitation, again producing references to the concept of mirror neurons. For example:

> they'd got to the point where, you know, they hadn't got much and what they had got was very limited and it was very clichéd ... they couldn't seem to generate ideas ... [but] they worked so much better when we showed that we were willing to generate ideas too.

There was a sense in which acting and generating in front of the children communicated both the types of mental processes required and their legitimacy:

> I can't do it wrong if I do what she's done ... so it's OK, I can take part in this now ... I can allow myself to be generative, even though people have told me I'm wrong before, this can't be wrong now.

Trainees spoke of there being transitions within a lesson, describing some lessons as 'like a sandwich' of thinking modes. They also discussed how transitions between dominant modes of thinking could sometimes be helpfully positioned at the boundary between lessons. Trainees also referred to instances when changing context and suspending evaluation had succeeded in dissipating fixated mind sets. Working with others was also seen as a valuable way of encouraging children to make links, including those links between interpretations of their own and others' ideas:

> but also working with other people and seeing what they do and taking your own interpretation of what they do – because they don't explain what they're doing and what they're saying – that, in turn, helps you generate ideas ... like with the Rorschach tests with the ink splots – what do you think you see? – you take your own interpretation and that helps you create your own mental links which puts you on further in the generative process.

Perhaps unsurprisingly, although the team had been at pains to point out that this was not the case, there remained a natural tendency for some trainees to assume a simple functional-anatomical mapping of cognitive processes, including those associated with generative and analytical modes of thinking:

> You're using almost two different parts of the brain there to do it, so like separating them into generative now and analytical at a different time ... so trying to switch.

Finally, the teacher trainees and their trainers were asked what they had got out of this experience of reflecting upon their practice in terms of psychological and neuropsychological concepts. First, there was a sense of having an improved theoretical understanding that supported existing practice, especially in terms of the role of 'warm ups'. Secondly, the trainees expressed a sense of being more empowered to intervene and support children's creative cognitive processes:

> so that when you go into the classroom, you can identify the different states, you know, that you can then manipulate or change it, and what's the point of that change. You as a teacher

can then change their way of thinking and make a more productive learning environment for your pupils.

Trainees referred to a number of issues influencing creativity that they felt provided insights into their own practice, and overall there appeared a new sense of responsibility for fostering abilities they had initially considered as entirely spontaneous and not amenable to teacher intervention:

> not all children/pupils/adults find it that easy to be creative, then when we go into schools, you can't just expect them to just improvise, just 'cos we can do it. It's up to us as teachers, then, to differentiate.

Issues regarding the difficulty in combining the language and perspective of natural science with educational thinking remained salient even in this final discussion, as some trainees struggled to find the appropriate terms by which to express their thoughts:

> Trainee: I think its reawakened (1) my curiosity, and (2) some previous revelations about environment and the effects that it has on people, and what they're capable of doing and how – and this is the only way I can think of saying it, how you can psychologically manipulate [laughter] – there's probably a better way to say it!
> Other [suggesting]: '. . . effect change?'
> Trainee: That's the one . . . [laughter], but you can look at and influence the environment and [thereby] people's way of thinking, and how to change that, and get the best out of people by doing that.

Conclusion

Overall, during this short intervention, the trainee teachers showed progression in their attention to, and understanding of, creative cognition in the classroom. This progression passed through stages that included:

(1) an initial high degree of enthusiasm.
(2) a flourishing of initial behavioural and conveniently prescriptive neuromyths.
(3) a daunting realisation that things were more complex and required attention to cognition.
(4) increase in meta-cognition, with neuroscience helping to 'biologise', 'concretise' and deepen concepts.
(5) emergence of concepts, language and reflective capability that allows deeper reflection, sensitivity and insights around personal practice in specific contexts, in terms of mind and brain.

Trainees' efforts to understand their own personal experiences of learning/creativity in terms of underlying cognitive processes appeared an important step in developing related insights into their teaching practice. Trainees sought to apply their new understanding in a variety of areas, including environmental effects and issues around the planning of activities such as the sequence of events and providing for individual differences. fMRI and other research involving imaging can be very effective in engaging non-specialists with thinking about the mind and the brain although, with this power to engage, also arise attendant dangers of encouraging myths such as simplistic phrenology. It was also found that neuroscientific case studies had a role in helping trainee teachers understand the mind and the brain, although their appropriateness as a more general teaching tool in the area of education may need further ethical consideration.

Here we have reported on an exploratory study that focused on the process by which pedagogical concepts can be co-constructed across neuroscience and education. We have not reported in any detail on the concepts developed (see Howard-Jones 2008) and these have not been formally evaluated. If such ideas are, as we hope, an improvement on the many 'brain-based' learning ideas presently being marketed, several issues will still need to be considered in determining their value, two of which deserve mentioning here. First, scientific knowledge of the mind and brain will always be partial and pedagogical ideas that draw on such knowledge will always require continuous updating and improvement. For example, the trainees were encouraged to use research findings to gain reflective insight into the creative behaviour of their pupils. However, the fMRI studies of 'normal' cognitive function presented to the trainees had been carried out with adults, whereas children's cognitive and neural processes may differ significantly from those of adults. As research on mind and brain progresses, these differences will inevitably need to be considered in terms of their pedagogical implications. Related to such considerations, trainees judged the understanding they had gained to be useful and it appeared to improve their ability to reflect on their practice, but its value in terms of improving practice still requires further investigation. We tentatively suggest that the concepts developed from a project such as ours could provide a helpful and stimulating contribution to teachers' systematic enquiries into their own practice. Such enquiries, which help develop teachers as reflective learners, are considered in themselves to be an important ingredient of effective teaching and learning (Hofkins 2007).

In our project, insights about mind and brain successfully highlighted a general message about how creativity involves a generative mode of thinking that is essentially different to the analytical mode predominant in school education. On the other hand, as was emphasised to the trainees, it is clear that individual creativity will always be a journey whose destination is unknown. Every creative journey is a unique experience, just as every brain is unique in terms of both its structure and functioning. For these reasons alone, neuroscience cannot entirely explain or demystify creative cognition and experience. However, using a process of co-construction that attends to both educational and scientific perspectives may produce new ways to think and talk about creativity and, in this way, help us to reflect upon the daily decisions we make as educators when fostering creativity in our students.

Acknowledgement

This research was made possible by a grant from the Education Subject Centre of the Higher Education Academy Network (ESCalate). The data presented, statements made and views expressed are solely those of the authors.

References

Bruer, J.T. 1997. Education and the brain: A bridge too far. *Educational Researcher* 26, no. 8: 4–16.
Chappell, K. 2007. The dilemmas of teaching for creativity: Insights from expert specialist dance teachers. *Thinking Skills and Creativity* 2, no. 1: 29–56.
Cohn, S. 2004. Increasing resolution, intensifying ambiguity: An ethnographic account of seeing life in brain. *Economy and Society* 33, no. 1: 52–76.
Cooper, R.B., and B. Jayatilaka. 2006. Group creativity: The effects of extrinsic, intrinsic, and obligation motivations. *Creativity Research Journal* 18, no. 2: 153–72.
Craft, A. 2000. *Creativity across the primary curriculum: Framing and developing practice.* London: Routledge.
———. 2006. Fostering creativity with wisdom. *Cambridge Journal of Education* 26, no. 3: 337–50.
Davis, A. 2004. The credentials of brain-based learning. *Journal of Philosophy of Education* 38, no. 1: 21–35.
Elliot, J. 1991. *Action research for educational change.* Buckingham: Open University Press.

Fitzgerald, K.D., R.C. Welsh, W.J. Gehring, J.L. Abelson, J.A. Himle, I. Liberzon, and S.F. Taylor. 2005. Error-related hyperactivity of the anterior cingulate cortex in obsessive-compulsive disorder. *Biological Psychiatry* 57: 287–94.

Forgays, D.G., and D.K. Forgays. 1992. Creativity enhancement through floatation isolation. *Journal of Environmental Psychology* 12: 329–35.

Geake, J.G. 2004. Cognitive neuroscience and education: Two-way traffic or one-way street?. *Westminster Studies in Education* 27, no. 1: 87–98.

Gehring, W.J., and R.T. Knight. 2000. Prefrontal-cingulate interactions in acion monitoring. *Nature Neuroscience* 3, no. 5: 516–20.

Hayes, D. 2004. Understanding creativity and its implications for schools. *Improving Schools* 7, no. 3: 279–86.

Hofkins, D. 2007. What's the evidence? In *Principles into practice: A teacher's guide to research evidence on teaching and learning*, ed. TLRP, 14–16. London: TLRP.

Howard-Jones, P.A. 2002. A dual-state model of creative cognition for supporting strategies that foster creativity in the classroom. *International Journal of Technology and Design Education* 12, no. 3: 215–26.

———. 2008. *Fostering creative thinking: Co-constructed insights from neuroscience and education.* Bristol: Escalate. http://escalate.ac.uk/4389.

Howard-Jones, P.A., S.J. Blakemore, E. Samuel, I.R. Summers, and G. Claxton. 2005. Semantic divergence and creative story generation: An fMRI investigation. *Cognitive Brain Research* 25: 240–50.

Howard-Jones, P.A., and S. Murray. 2003. Ideational productivity, focus of attention and context. *Creativity Research Journal* 15, no. 2/3: 153–66.

Howard-Jones, P.A., J. Taylor, and L. Sutton. 2002. The effects of play on the creativity of young children. *Early Child Development and Care* 172, no. 4: 323–8.

Kris, E. 1952. *Psychoanalytic explorations in art.* New York: International Universities Press.

Lythgoe, M.F.X., T.A. Pollak, M. Kalmus, M. de Haan, and W. Khean Chong. 2005. Obsessive, prolific artistic output following subarachnoid haemorrhage. *Neurology* 64: 397–8.

National Advisory Committee on Creative and Cultural Education (NACCCE). 1999. *All our futures: Creativity, culture and education.* London: Department for Education and Employment.

Pickering, S.J., and P.A. Howard-Jones. 2007. Educators' views of the role of neuroscience in education: A study of UK and international perspectives. *Mind, Brain and Education* 1, no. 3: 109–13.

Rizzolatti, G., L. Fadiga, L. Fogassi, and V. Gallese. 2002. From mirror neurons to imitation: Facts and speculations. In *The imitative mind: Development, evolution, and brain bases*, ed. A.N. Meltzoff and W. Prinz, 247–66. New York: Cambridge University Press.

Schumacher, R. 2007. The brain is not enough: Potential and limits in integrating neuroscience and pedagogy. *Analyse and Kritik* 29, no. 1: 38–46.

Wallas, G. 1926. *The art of thought.* New York: Harcourt, Brace & World.

Werner, H. 1948. *Comparative psychology of mental development.* New York: International Universities Press.

Wundt, W. 1896. *Lectures on human and animal psychology.* New York: Macmillan.

Index